The Am Fire Phone

Master your Amazon smartphone
including Firefly, Mayday, Prime, and all the top apps

Scott McNulty

The Amazon Fire Phone
Master your Amazon smartphone including Firefly, Mayday, Prime, and all the top apps
Scott McNulty

Peachpit Press
Find us on the web at: www.peachpit.com
To report errors, please send a note to errata@peachpit.com

Peachpit Press is a division of Pearson Education

Copyright © 2015 by Scott McNulty

Editors: Clifford Colby and Scout Festa
Compositor: Danielle Foster
Indexer: Valerie Haynes Perry
Cover Designer: Aren Straiger

Notice of Rights
All rights reserved. No part of this book may be reproduced or transmitted in any form by any means, electronic, mechanical, photocopying, recording, or otherwise, without the prior written permission of the publisher. For information on getting permission for reprints and excerpts, contact permissions@peachpit.com.

Notice of Liability
The information in this book is distributed on an "As Is" basis without warranty. While every precaution has been taken in the preparation of the book, neither the author nor Peachpit shall have any liability to any person or entity with respect to any loss or damage caused or alleged to be caused directly or indirectly by the instructions contained in this book or by the computer software and hardware products described in it.

Trademarks
Many of the designations used by manufacturers and sellers to distinguish their products are claimed as trademarks. Where those designations appear in this book, and Peachpit was aware of a trademark claim, the designations appear as requested by the owner of the trademark. All other product names and services identified throughout this book are used in editorial fashion only and for the benefit of such companies with no intention of infringement of the trademark. No such use, or the use of any trade name, is intended to convey endorsement or other affiliation with this book.

ISBN 13: 978-0-134-02289-5
ISBN 10: 0-134-02289-0

10 9 8 7 6 5 4 3 2 1

Printed and bound in the United States of America

For my mom.

Acknowledgments

The author gets to have their name on the book, but a lot of people helped to make this possible. The Peachpit staff continues to impress. Cliff Colby, despite his own good sense, is always onboard when I suggest a new book. Scout Festa did an amazing job, on a tight deadline, of shaping my words into a book. My thanks also to Dennis Fitzgerald for managing the production of the book and to Liz Welch for copyediting.

Any mistakes are my own.

Contents

CHAPTER 1 Introduction — 1
   What's Not Included? — 3
   Using This Book — 4

CHAPTER 2 Getting to Know Your Fire Phone — 5
   Setting Up Your Fire Phone — 9
   The Fire Phone Interface — 12
      Panels — 14
      Quick Settings — 17
      Home grid — 21
      Quick Switch — 26

CHAPTER 3 Working with Text — 27
   Keyboard — 28
      Predictive text — 28
      Spellcheck — 31
      Trace typing — 31
      Speech to text — 32
      Advanced keyboard — 33
   Selecting Text — 35
      Inserting the cursor — 37

CHAPTER 4 Email — 39
   Setting Up an Email Account — 40
   Deleting Accounts — 44
   Account Settings — 46
      Display — 46
      Sync and data settings — 47
   Checking and Reading Your Email — 49
   Composing Email — 55
   Managing Email — 58
   Customizing Your Mail — 60

CHAPTER 5 Contacts 65
    Getting Started 66
    Transferring Contacts from Your Old Phone to the Fire. 67
    Setting Up a Profile and Contacts 68
        Navigating contacts. 72
        VIPs. 73
    Settings 76

CHAPTER 6 Messaging 79
    Sending a Text. 82
    Reading Texts 84
        Find a message 85
        Notifications 85

CHAPTER 7 Phone 87
    Making a Call. 88
        Conference call 90
        Answering a call 90
    Settings 92
        Volume 93
        Additional settings. 93

CHAPTER 8 Calendar 95
    Setting Up Your Calendar 96
        Facebook 96
    Viewing Your Calendar 97
    Events 100
        The right panel 103
        Creating an event 104
        Responding to an event invitation 106
    Settings and notifications 107

CHAPTER 9 Silk 109

    Visiting a Webpage 110

        Reader mode 117

        Tabs 119

    Navigation 121

        Bookmarks 121

        Saved Pages 123

        Trending Now 124

        Downloads 125

        History. 126

        Settings. 126

CHAPTER 10 Apps 131

    Amazon Appstore 132

        The apps. 132

        Buying and installing 135

        Test Drive before you buy 136

        In-app purchases 138

        Buying Amazon Coins 138

        Navigating the Appstore 139

        Left panel 140

        Sideloading. 142

        Appstore settings 146

    Stock Apps 147

        Games. 147

        Maps. 149

        Clock. 153

        Calculator 155

CHAPTER 11 Camera and Firefly — 157

Taking a Picture — 158
   Taking a panorama — 159
   Lenticular photography — 160
   HDR — 162
Shooting a Video — 162
Pictures — 163
Firefly — 166

CHAPTER 12 Kindle — 171

Kindle Basics — 172
   Reading — 175
Kindle Store — 180
Newsstand — 182
Docs — 183

CHAPTER 13 Music — 185

Playing Music — 186
Playlists — 188
   Managing Music — 190
Purchasing Music — 191

CHAPTER 14 Instant Video — 193

Renting/Purchasing/Downloading — 194
Watching — 196
   Second screen — 198
   Watchlist — 199
Your Own Videos — 199
Video Settings — 199

CHAPTER 15 Voice Control   **201**

    Make Your Fire Listen 202

    Phone Calls 202

    Messaging 203

    Searching. 204

CHAPTER 16 Amazon Prime   **205**

    Shipping 206

        Share the shipping. 206

    Prime Instant Video 207

    Prime Music. 207

    Kindle Lending Library 209

    Kindle First. 210

CHAPTER 17 Security   **211**

    On Your Fire 212

        Encryption. 213

        Updates 215

    Amazon.com 216

        Deregister 216

        Locate on map 217

        Remote Lock, Factory Reset, Remote Alarm. 218

CHAPTER 18 Help   **219**

    Mayday 220

    Self Service 221

    Amazon.com help 222

    Index 223

CHAPTER 1

Introduction

Purchasing a smartphone is something that one shouldn't do lightly. Carriers—AT&T, Verizon, and Sprint, to name a few—subsidize most smartphone purchases in America. This means that in exchange for signing a contract for a certain number of years of service, the carriers knock hundreds of dollars off the cost of the phone. They make this up month by month as you pay your bill for service.

Why do I mention this? Because a two-year contract is a commitment, and the Fire phone is a gamble. It is Amazon's first entry into the smartphone market, so who knows if you'll like it? And if you don't, you'll have to pay a hefty fee to cancel your contract, or buy a replacement phone at full price.

I assume that since you're reading this book you either have a Fire phone or are interested in one. Either way, you should be aware of some of the unique features of the Fire phone, which will be covered in detail in this book.

As with all smartphones, the Fire phone makes phone calls, sends text messages (and MMSs), comes with a web browser (Silk), has a camera, and is able to connect to your email and calendar. The Fire phone can also access thousands of apps via the Amazon Appstore, not to mention all the videos, Kindle books, and music that Amazon sells.

In addition to the standard suite of features, the Fire phone offers up a few unique things:

- **Dynamic Perspective.** This is probably the defining feature of the Fire phone, and a little difficult to describe in words. Dynamic Perspective is definitely one of those "you have to see it to truly get it" features. It works like this: the Fire phone has a number of sensors that know where your face is as you're looking at it (yes, even in the dark). On the surface that might seem creepy, but it allows the Fire to do some very clever things. As you move and tilt the phone, the display reacts by revealing some new information in an app, or displaying a new section of the scene you're looking at as if you're looking through a window. Dynamic Perspective is used throughout the Fire phone's interface, and Amazon has made it available to people who make apps for the phone, so it'll start popping up in third-party apps as well.

- **Firefly.** Cynics opine that Amazon's devices (eInk Kindles, the Fire tablets, the Fire TV, and the Fire phone) are designed only to make it easier for you to spend money on Amazon goods and services. Clearly, Amazon would like you to buy everything from them, and Firefly makes that very easy indeed. Firefly transforms your Fire's camera into a scanner of everyday objects, barcodes, music, videos, and even URLs. Fire up Firefly, point it at something, like a book, and chances are the Fire will identify it and display some information about how you can purchase it on Amazon.com. Much as with Dynamic Perspective, Amazon has opened the Firefly technology to developers, so look for integration with third-party apps that allow you to do things like point your phone at a package of food and have the Fire display nutritional information about the item.

- **Mayday.** You had the good sense to purchase this book, but even still you might come across a feature on your Fire phone that you can't get to work correctly, or that you have some questions about. With Mayday you simply tap a button and you're connected to a real live Amazon support person who can help you with any phone-related issue you may be having.

- **Unlimited photo storage.** The Fire's camera takes some impressive photos, and you don't have to worry about backing them up. They are all automatically synced to your Amazon Drive, which is hosted on Amazon's servers. You can store an unlimited number of photos taken with your Fire on this storage—so shoot away!

- **Amazon integration.** Amazon makes this phone, so it should come as no surprise that Amazon services work well with it. You can purchase Kindle books, Amazon MP3s, Amazon videos, and even physical items from Amazon.com right from the Fire. The Fire is even aware of when your items ship from Amazon and are delivered and will alert you (which is great if you're as impatient as I am). If you purchase your Fire phone from Amazon.com, it comes preconfigured with your Amazon account, so setup is even easier.

- **A year of Amazon Prime.** For a limited time—though Amazon won't say how limited that time is—every Fire phone comes with a year of Amazon Prime. Prime, which started off as a shipping program, has blossomed into a service that offers a variety of benefits and usually costs $99 a year. Prime members get free 2-day shipping (and cheap 1-day shipping) on eligible items from Amazon.com, can stream unlimited videos from Prime video, can listen to music via Amazon's Prime Music service, and can borrow one ebook from the Kindle Lending Library a month. And that's just the tip of the Amazon Prime iceberg. Prime is fully covered in Chapter 15.

▶ **TIP** If you're already a Prime subscriber, your membership is automatically extended for a year when you purchase a Fire phone—for a limited time.

What's Not Included?

The Fire phone offers up a number of great features in addition to those you normally find on an Android smartphone. That's right, the Fire phone is running Google's Android operating system. Amazon has greatly customized it, but at its core the Fire phone is an Android device.

Why should you care about this? For most people it won't matter at all, but some folks might assume that the Fire phone running Android means that you'll have access to all the apps available in the Google Play Store (that's the Android app store) and to Google's own suite of apps, such as Google Maps.

This is not the case. Google apps aren't installed on the Fire by default, and they aren't available in the Amazon Appstore. You can connect your Gmail account to your Fire phone and send email and check your calendar, but you won't have that deep integration with Google services that regular Android phones offer.

> **TIP** There is a way to install apps from Google's App Store onto your Fire phone. It's covered in the "Sideloading" section of Chapter 10.

Using This Book

This book doesn't assume that the reader has any knowledge of the Fire phone or of smartphones in general. The Fire is a touch device, and I will be using some terms to describe what you should do with your device:

- Tap: Using one finger, quickly press and release on the Fire's screen.
- Long tap: Press one finger against the phone for a couple of seconds before releasing.
- Double tap: Rapidly tap twice with one finger.
- Pinch: Use two fingers to pinch in and out on the screen. This zooms in and out.
- Swipe: Pressing your finger against the phone, move it up or down on the screen, releasing as you reach the apex of the swipe.
- Tilt: Move one of the Fire's edges toward you while moving the other away.

All the instructions in this book assume that the bottom edge of the Fire phone's screen is the edge nearest the home button (the only button on the front of the Fire).

Additionally, if you need to tap a number of things in sequence I'll denote that with a >. For example, if you need to open Quick Settings, tap the Settings icon, and then tap Display, that will look like this: Quick Settings > Settings > Display.

CHAPTER 2

Getting to Know Your Fire Phone

Your Fire phone has arrived and you've ripped it out of the packaging. Included with the phone are headphones (which are magnetized to keep the earbuds together), a charger, and a cable.

Before you turn on the phone, let's take a quick look at its physical features. Both the front and back of the phone are made out of something called Gorilla Glass, which is just like glass only tougher. It is highly scratch resistant and resilient, though I suggest you avoid dropping the phone on hard surfaces.

▶ **TIP** Amazon sells a wide variety of cases for your Fire phone. You might want to look into one if you have a tendency to drop things, or just want your phone to stand out in the crowd.

The sides of the Fire are made of a rubbery plastic that gives your hand something to grip onto. This is great because the glass can be a bit slick.

Almost the entire front of the Fire is composed of the 4.7-inch multitouch display. The screen is high definition (720p at 315 pixels per inch). Surrounding the display are five circles. The circles in the corners are Dynamic Perspective sensors, which track where your face is as you use the phone. The remaining circle, to the left of the Dynamic Perspective sensor in the upper-right corner of the phone, is the front-facing camera.

There aren't a lot of physical buttons on the Fire (2.1):

- Home button: The only button on the front of the Fire, the home button brings you to the home screen no matter where you are in the Fire interface and launches Quick Switch (more on that in the "Quick Switch" section of this chapter).

- Volume buttons: On the left edge of the Fire, these two buttons are volume controls (top turns up the volume, bottom turns it down).

- Camera button: Press this button and the camera app launches. When the camera app is launched, pressing this button again makes your Fire take a picture (or video, depending on the settings in the camera app).

▶ **TIP** If you press and hold the camera button, Firefly launches.

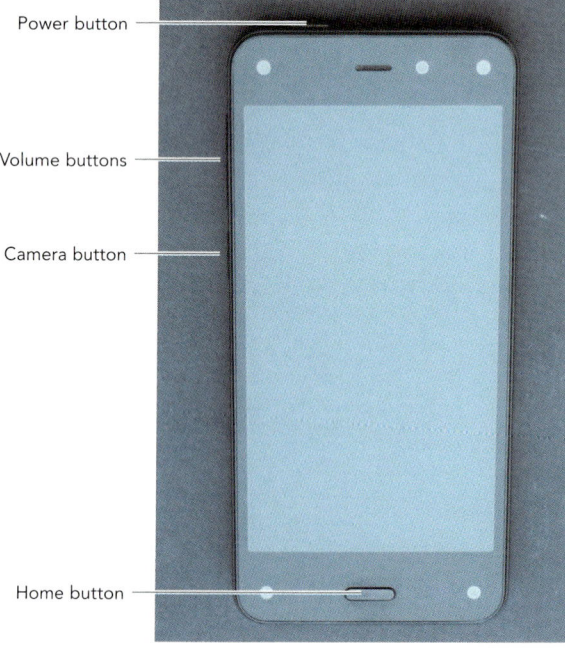

2.1 The front of the Fire phone has only one button.

- Power button: At the top left edge of the Fire is the power button. When the phone is off, pressing and holding this button turns it on. When the phone is powered on, quickly pressing this button puts the phone to sleep (the display turns off to save power). Press any of the buttons on the phone to wake it up from sleep.

To turn off your Fire phone, press and hold the power button. This brings up a couple of options (2.2):

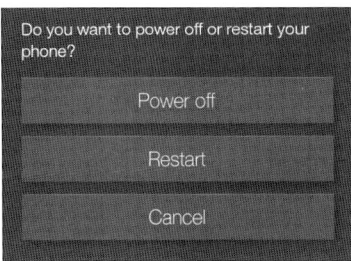

2.2 Long pressing the power button brings up the shutdown options.

- Power off turns off the phone.
- Restart turns off the phone and automatically turns it on again (a great troubleshooting step).
- Cancel is if you have decided you don't want to turn off your Fire after all.

Those are all the buttons on the Fire, but there are a few more things to be aware of:

- Headphone jack: Located on the top right edge of your Fire, this is where you plug in the included (or any other) headphones. This is a standard headphone jack, so it'll work with any accessories (speakers or the like) that plug into a headphone jack.
- Speakers: At the top and the bottom edges of the phone you'll see bunch of holes clustered together. These are the two speakers of the Fire, and since there are two speakers, the Fire phone is able to play audio in stereo (and it doesn't sound too shabby considering the size of the speakers).
- Microphone: The Fire is a phone, so it needs a microphone to pick up your voice. This microphone is disabled when you plug in headphones with an integrated mic (like the headphones that are included with the Fire).

- SIM tray: Your Fire needs a SIM (subscriber identification module) card in order to connect to your carrier's cell network. This card slides into the Fire in a tray on the left edge of the phone. You'll probably never need to remove or replace the SIM, but if you do just straighten a paperclip and poke it into the hole on the tray. The tray will pop open with your SIM in it.
- Micro USB port: At the bottom of the Fire is a Micro USB port used for charging. Plug in the provided cable to charge your phone or to connect it to a computer (more on this in Chapter 10).

That covers the front of the Fire. Flipping the Fire over reveals a shiny silver Amazon logo as well as some regularity information at the bottom of the Fire's back (2.3).

In the upper-right corner of the Fire's back you'll see:

- 13 MP rear-facing camera: This is the camera that the Fire defaults to using whenever you start the included camera app.
- LED flash: The Fire will automatically use the flash when you're trying to take a picture in a low-light situation. You can also use the flash as a flashlight.

2.3 The back of the Fire phone.

Setting Up Your Fire Phone

1. To turn on your Fire phone for the first time, press the power button at the top left of the phone. The Amazon logo will appear, quickly followed by a Fire logo that pulses as the phone boots up.

2. Select the language you'd like the interface to be in (this book assumes you'll go with the default: English), and tap Next (2.4).

3. If you purchased your phone directly from Amazon.com, it is already registered to your Amazon account. If you purchased elsewhere, you need to enter your Amazon account details (or create an Amazon account by tapping "Start here") before you can start using your phone (2.5).

4. Once you've entered your Amazon account's information, tap Register. The Fire registers to your account.

▶ **NOTE** You can skip registering your phone with your Amazon account by tapping "Register later," but that will only delay the inevitable. The Fire isn't much fun to use if it isn't registered to an Amazon account.

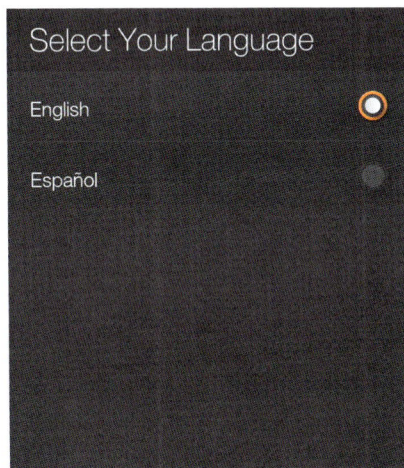

2.4 You have two language options when setting up your Fire.

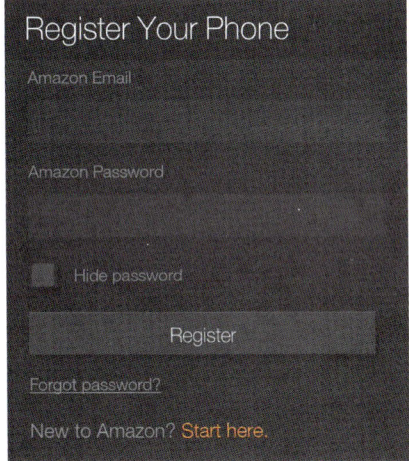

2.5 Use your Amazon account to register your Fire phone.

5. Now you must confirm that the Fire is registered to the correct Amazon account (this step happens on pre-registered Fires as well as on manually registered devices) (2.6). If this isn't the account you want to use, tap "Not [your name]?" to change accounts. Otherwise, tap Next, keeping in mind that by tapping Next you're agreeing to Amazon's terms of service, which you can read by tapping "all the terms found here."

6. The Fire scans for Wi-Fi networks and lists the networks it finds (2.7). Tap a network to join it (you'll be prompted for the Wi-Fi password if one is needed). If you don't see your Wi-Fi network listed, you might need to manually add it by tapping Add Network. Fill in your Wi-Fi network's details, and tap Join.

    "Connected" will display under the network name when you've successfully joined it (2.8).

    Tapping No Thanks will progress to the next step without joining a Wi-Fi network.

7. Tap Next to get to the Enable Location Services section. If you tap Enable, your Fire phone will enable GPS for apps that work best when they know where you are (Maps, for example). If you don't want to enable GPS, tap No Thanks.

2.6 Amazon wants to make sure you registered with the correct account.

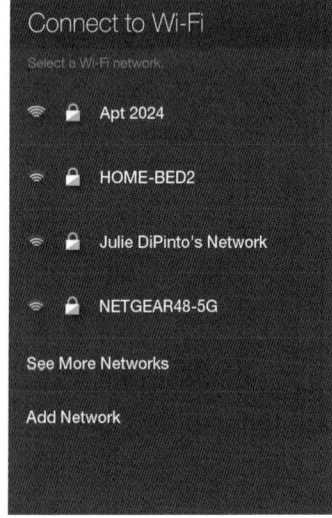

2.7 The list of available Wi-Fi networks.

2.8 When you successfully connect, the word *Connected* appears in orange beneath the Wi-Fi network name.

8. Next up is enabling backups. Your Fire can automatically back itself up to Amazon's server. This backup is free and includes settings, notes, search history, bookmarks, and more. I suggest that you enable this setting just in case you ever need to reset your Fire to factory settings, or need to register a replacement Fire. If you've reset your Fire, you can restore from a previously taken backup, as in 2.9.

9. You can connect your Fire to your Twitter and Facebook accounts, which enables you to share items from your phone to your social networks (2.10). Tap either Twitter or Facebook, and enter the appropriate credentials to link your phone to that account.

10. Tap Next when you're done, and you're offered a 30-day trial membership in Audible, Amazon's audio book service. The "Start your 30-day free trial" button is large and tappable at the bottom of the screen. If you aren't interested, tap the smaller "No thanks" text above the button.

11. You're almost done setting up your new Fire. You can now watch a 5-minute tutorial video by tapping Start. I recommend you watch it, though this book covers everything in the video and more. If you aren't into watching videos and you want to dive into your Fire, tap Skip.

Now your phone is ready for you!

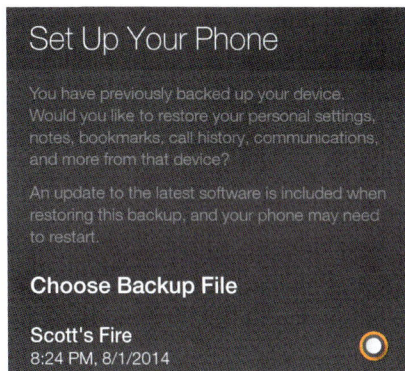

2.9 When a backup is available, you can restore from it.

2.10 Connect Twitter or Facebook to your Fire phone.

The Fire Phone Interface

The Fire's home screen is where you'll access recently used apps and media (2.11). The home screen's largest feature is the Carousel. That's the row of large icons at the top of the home screen. Swipe to the left to look through the Carousel (and right to go back). As you start to use apps, you'll notice that the first item in the Carousel will always be your most recently used app (or the most recently viewed piece of media).

Under each icon on the Carousel, additional related items are displayed. Some applications, like the Appstore for example, use this space to give you a preview of the content in the app (2.12). Some apps aren't written to give you additional information on the Carousel, so you'll see suggested apps from the Appstore that you might like.

To launch an app from the Carousel, just tap its icon. Press the home button to return to the Carousel.

Long tapping an icon in the Carousel will reveal a menu with different options, depending on what the icon represents (2.13).

2.11 The Carousel shows your recently used apps and media.

2.12 The Carousel displays related items below the main icons.

2.13 Long tap an icon to display these options.

Pin to Front pins that item to the front of your Carousel. It won't budge from that spot no matter which apps or media you open on the phone. While pinned icons are displayed at the very front of the Carousel, they aren't the first icon displayed when you are on the home screen. The Carousel still displays the most recently used items front and center, but with the pinned icons a left swipe away.

Pinning an app is great way to have quick access to a favorite app or to that book you're currently reading. Keep in mind that you can pin more than one thing to the front of the Carousel.

All apps pinned to the front of the Carousel display a little pin icon under them (2.14). Long tapping a pinned icon brings up a menu with a couple of unique options. Unpin from Carousel returns the selected icon back to normal Carousel behavior. If you have more than one thing pinned to the front of your Carousel, there is an obvious problem: only one thing can be at the start of the Carousel. Pinned items are arranged in the order that they were pinned. If you want to move one pinned item to the front, long tap it and then tap Move to Front.

2.14 Long tap a pinned icon to unpin it.

The Remove from Carousel option will always be included, and tapping it does what you think it does: it removes the app or media icon from the Carousel. This does not remove the item from your phone; it is just removed from the Carousel.

If you long tap an app that's installed on the Fire (more on installing apps in Chapter 10) or a media file you've played/read/watched, you'll see a Remove from This Device option. Tapping this immediately removes that item from your Fire. There isn't a confirmation pop-up, so make sure you're certain when removing things from your device.

Finally, if you long tap a book, video, or song in the Carousel, you'll see one additional option: Pin to Home Grid. Tapping this adds a shortcut to that piece of media to the home grid, which is usually limited to displaying apps (see the "Home grid" section later in this chapter for more information).

At the bottom of the home screen, you'll see a row of four icons. By default, these four icons are Phone, Messaging, Email, and Silk Browser. These icons act as a quick launcher for your frequently used apps, though unlike the Carousel, these four icons are static. They only change if you change them. Tap any icon to launch its app. You can remove apps from and add apps to this row, which we'll cover later in this chapter.

Peek

If you've used a smartphone before, you might notice that something is missing from the home screen: the status bar. That's the black strip along the top of the display that includes the time, the signal strength, and other pieces of information. That status bar is there; it is just hidden thanks to Dynamic Perspective (see Chapter 1 for more information).

If you move your Fire to the right or left slightly while on the home screen, you'll see the status bar appear as if by magic. This is a feature that Amazon calls Peek. Most areas of the phone support this, so be sure to move the phone a little as you explore so you can see all the information that the Fire has to offer. To make the status bar visible all the time, go to Settings > Display > Show Status Bar, and set it to On (see the "Quick Settings" section of this chapter for more information).

Panels

Panels are another important bit of user interface on your Fire phone. There are two panels on almost every screen in the Fire interface, including the home screen: the right panel and the left panel. The content of each panel is contextual, meaning that each app on your phone displays different things in each panel. The left panel is home to app-specific navigational options. For example, you can go to playlists, artists, or songs via the left panel of the Music app. The right panel contains something that Amazon calls "delighters." Although I don't like that name, the concept is sound: the right panel contains extra bits of functionality that enhance the app you're using. Staying with the Music app for a moment, the right panel displays lyrics of the song you're listening to (assuming the song you're listening to supports this feature). Delightful!

No matter which panel you want to interact with, the methods of revealing the panels are the same across the Fire's screens. That's right, *methods*: there are actually two ways to reveal either the right or left panel. The first way is the traditional swipe method. Swipe your finger from the left edge to the right to reveal an app's left panel. Do the same from right to left to summon the right panel. To dismiss either panel, swipe the opposite way.

The traditional method works, but it does require two hands (most of the time)—one to hold the Fire and the other to swipe. What if you're struggling with a bag in one hand but want to access a panel on your Fire in the other? That's where one-handed gestures enter the picture.

To reveal the left panel using only one hand, tilt the left edge of the Fire toward you. This will cause the left panel to slide out. To dismiss it, tilt the right edge toward you. Reverse the process for the right panel.

The home screen left panel

The home screen left panel contains a list of shortcuts to various parts of the Fire interface (2.15):

- Apps: the Appstore
- Games: the Games section of the Appstore
- Web: launches the Silk browser
- Music: the Music Store
- Videos: instant video
- Photos: your photos
- Books: your books
- Newsstand: your magazines
- Audiobooks: your audiobooks
- Docs: your personal documents (more about these in Chapter 11)
- Shop: the Amazon Shopping app
- Prime: a list of all your Amazon Prime benefits (if you're a subscriber) or a way to sign up for Amazon Prime if you aren't already signed up

2.15 The left panel on the home screen has shortcuts to different sections on your phone.

Tapping any item in the left panel brings you right to that area of the Fire phone. Using Peek, you can see a little more information about some of the items.

The home screen right panel

The right panel on the home screen contains a variety of information gathered from around your phone (2.16). At the very top of the panel, the local weather is displayed (assuming you've enabled Location Services; otherwise, the phone has no way of knowing where it is).

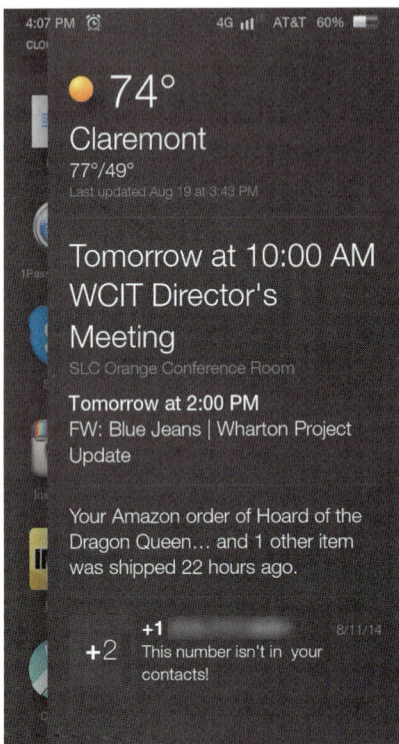

2.16 The right panel displays information from across the phone, such as the current weather.

Tapping the weather launches the full Weather app so you can see the forecast for the week, and not just the current conditions.

Right below the weather, upcoming appointments and recently received emails are displayed. Tap one of the entries to jump right to the full details of the event or email.

Under the email and calendar notifications is one of my favorite features of the home screen right panel: the Amazon order status. When you place an order for a physical object on Amazon.com, using the same account your Fire is registered with, the order details show up here. As the order progresses through the packing and shipment process, the entry on the right panel updates, including a notification when your Amazon package is delivered. If you tap the order in the right panel, you're taken to the order details in the Amazon Shopping app. There, you can cancel the order (if the order is still cancelable) or find the full tracking options for packages that have been shipped (this requires entering your Amazon password).

Right below the text/call information, you'll see recent phone calls and received text messages, along with the name and picture of the person who sent them (if they are in your Contacts, which is covered in Chapter 5). Tap a phone number to launch the Phone app and return the call (don't worry, you'll still need to tap the Call button, so there isn't much chance of accidentally returning a call), or tap a text message to reply to it.

Quick Settings

The third, and final, panel available in the Fire interface is Quick Settings (2.17). Accessible from anywhere in the Fire interface, this panel gives you fast access to a handful of settings that you can quickly toggle on and off or adjust, without interrupting what you were doing.

2.17 The top eight icons give you ready access to settings.

CHAPTER 2: GETTING TO KNOW YOUR FIRE PHONE   17

There are two ways to access the Quick Settings: swipe your finger from the top of the Fire downward and the Quick Settings panel slides down. You can also flick your phone downward and then back up quickly (though hold on tight, you don't want to throw your phone across the room!) to call it up.

To dismiss the Quick Settings panel, swipe up with your finger or flick your wrist once more.

The Quick Settings panel has eight icons and a slider. If you use Peek, labels for each icon appear. Here's what they do:

- Airplane Mode: If you're been on an airplane any time in recent memory, you know the drill: you have to turn off all electronic devices during takeoff and landing, and turn off your cellphone for the entire trip. You can certainly turn off your Fire by holding down the power button, but Airplane Mode allows you to keep using the phone without potentially causing issues to the plane. Tap the Airplane Mode icon; it turns orange, indicating that it is active. All the wireless capabilities of the phone are disabled (Bluetooth, Wi-Fi, and cellular), and an airplane icon appears in the status bar (2.18).

2.18 The plane icon signifies that your phone is in Airplane Mode.

- Wi-Fi: Tap the Wi-Fi icon to toggle Wi-Fi on (the icon is orange) or off (a white icon). When Wi-Fi is on and you're in range of a Wi-Fi network you've previously connected to, the phone will automatically connect to that network again. If you haven't connected to any of the in-range networks, you'll need to manually connect to your network of choice.

    Long tap the Wi-Fi icon and you're taken to the Wi-Fi settings screen (2.19). You'll see a list of the Wi-Fi networks your Fire can see. Those networks with a lock icon require a password to join.

    Tap a network to join it. If it is password protected, you'll be asked for the password before you're allowed to join the network. As soon as you connect, you'll see the familiar Wi-Fi strength icon displayed in the status bar . The more arcs, the stronger the connection.

    When you're connected, the icon is labeled with the name of the wireless network you're on. You can see it if you Peek at the Quick Access panel.

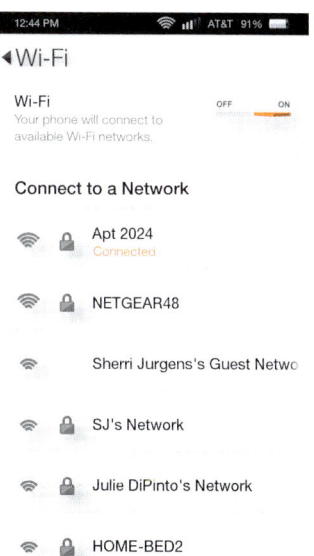

2.19 A lock icon indicates that the network is password protected.

▶ **TIP** You can turn on Wi-Fi when your Fire is in Airplane Mode by bringing up the Quick Settings panel and tapping the Wi-Fi icon. Now you can enjoy your flight's Wi-Fi without turning on your phone's cellular radio.

- Bluetooth: You can toggle on and off your Fire's Bluetooth by tapping the Bluetooth icon. Long tap the icon to access the Bluetooth settings.

- Flashlight: The Fire has a built-in LED flash so you can take photos in dark places. Tapping the Flashlight icon turns that flash on and keeps it on so you can use your $600 phone like a flashlight on your way to the bathroom late at night. Tapping the icon again turns off the light.

- Sync: Your phone is registered to your Amazon account. You can use that Amazon account to purchase things from Amazon.com and have those things appear on your Fire phone. Your Fire phone will regularly check to see if there is any content that it needs to sync to your phone, but tapping the Sync icon forces the Fire phone to check at that moment and pull down any new content (like Kindle books) to your phone. The icon spins around as it syncing.

- Settings: Tap this icon to access the Settings screen of your Fire. Throughout this book we'll be dipping into the settings to change some things.

- Mayday: Tap the Mayday icon and you'll be taken to a confirmation screen explaining that you're about to be connected to a live video session with an Amazon tech support person (2.20). If you haven't tapped Mayday inadvertently, tap the Connect button to complete the Mayday connection. Mayday is covered in detail in Chapter 17.

- Search: As you use your Fire more and more, you'll accumulate media, apps, and messages. Finding something that you're looking for might become a little difficult. That's where Search comes in. Tap the magnifying glass icon and a search box appears; it will list previous searches, if there are any (2.21).

  Type in whatever you're looking for, and Search will look across all the apps, media, and settings on your phone, as well as list suggested web searches and Amazon store searches. Tap an item to access it, or tap a suggested web or store search to launch that search in the appropriate app.

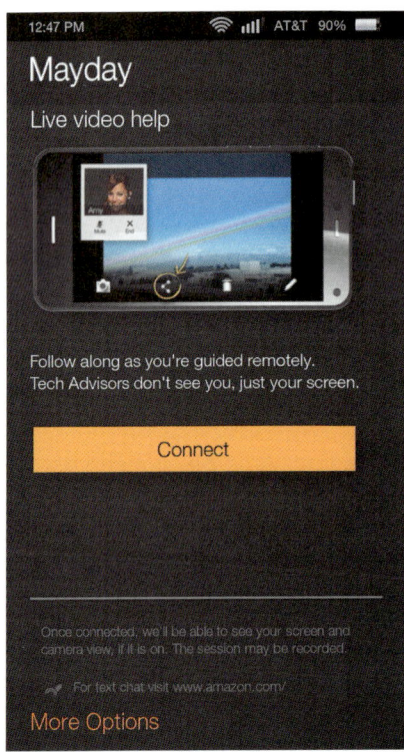

2.20 Mayday is live video help from friendly Amazon employees.

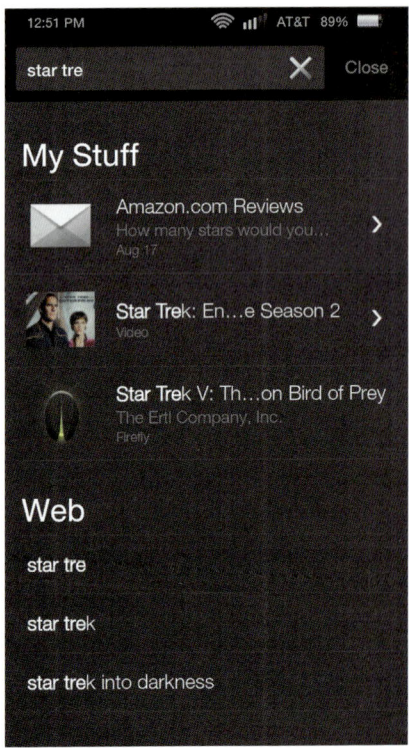

2.21 Search looks for items across your phone.

20 THE AMAZON FIRE PHONE

Under the row of Quick Settings icons, you'll find a slider that sets the brightness of your Fire's display. Slide it from left to right to increase the display's brightness; slide from right to left to decrease the brightness. Keep in mind that the brighter the display, the less battery time you'll have (the display consumes the most energy of anything on your Fire).

Right next to the slider you'll see an Auto button. When Auto brightness is enabled, the line above Auto will be orange. If it is silver, tap it to enable Auto brightness and the screen will automatically dim or brighten. If you decide you'd rather control the brightness yourself, simply slide the brightness slider left or right. This automatically disables Auto brightness.

▶ **NOTE** There's one more section of the Quick Settings panel: Notifications, which is covered in the next chapter.

Home grid

The Carousel is a great way to see all the apps and media that you've used recently, but how are you supposed to see *all* your apps?

The home grid shows all the apps you have installed on your Fire (2.22).

2.22 The home grid lists apps both on your Device and in the Cloud.

CHAPTER 2: GETTING TO KNOW YOUR FIRE PHONE   21

There are two ways to get to the home grid (noticing a trend)? Swipe your finger up from the bottom of the Fire while you're on the home screen, and the home grid slides up. On the Carousel, you can get to the home grid by pressing the home button. This means you can jump to the home grid no matter where you are in your Fire by pressing the home button to return to the Carousel and then pressing it again.

The home grid is composed of as many pages of grids as you need to display all the apps on your phone. As you swipe up on the home grid, dots appear on the right edge of the home grid. Each dot represents a page of apps, with the position of the page you're currently on represented by a white dot amid gray dots. The position of the white dot tells you at a glance how many pages of apps are available above and below your current page.

To switch pages, swipe up or down. Tap any app icon to launch it.

The home grid has two tabs: Device and Cloud. At the upper-left corner of the home grid, a button allows you to switch tabs. The current view is the one with the orange underneath it (2.22 shows the Device tab active). The Device tab lists all the apps you have installed on your Fire. Tap Cloud to see all the apps you've purchased from Amazon.com (2.23).

Any apps on this screen that are installed on your device will have a little checkmark on them (like 1Password Reader in 2.23). Tap an uncheckmarked app to install it on your Fire. An orange progress bar displays as the app downloads, and you'll be notified once it has been installed. That app will then show up on both the Device and Cloud tabs. Tapping it in either view will launch the app.

Arranging your apps

You can arrange the apps on your home grid in whatever manner you like. Long tap an icon that you'd like to move. The Fire will vibrate briefly to let you know you've selected an icon. Without lifting your finger from the display, drag the icon to where you would like it (2.24). The other app icons will move out of the way. Once you've found a new place for it, lift your finger. The icon stays where you placed it and the others arrange themselves accordingly.

To move an app from one page to another, use the same method to select the icon and then move it either to the top of the page (to move it one page up) or to the bottom of the page (one page down). If you are on the last page of your home grid and move an app to the bottom of the page, a new page will be created.

Note that the first row of icons on the first page of the home grid is displayed on the home screen under the Carousel. By default, Phone, Messaging, Silk Browser, and Email are in that row but you can put any four apps there, in any order. Use the same method to move an icon into that row.

▶ **TIP** You can also move icons out of that row, if you don't want four icons to be displayed.

As you get more and more apps on your Fire, you might want to group them. You can create *collections*, what other operating systems call folders, of apps on the home grid. Long tap one of the apps you want in the collection, and drag it over another app you want in the collection (collections must initially contain at least two items, but once created there is no such restriction) (2.25).

2.23 The Cloud tab of the Home grid.

2.24 Long tap an icon and drag it to a new place.

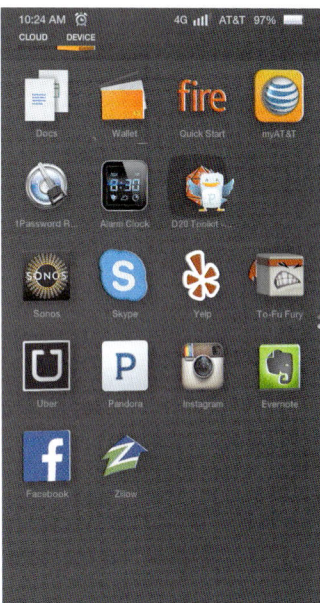

2.25 Drag an icon on top of another to create a collection.

CHAPTER 2: GETTING TO KNOW YOUR FIRE PHONE   23

You're prompted to name the collection (2.26). Keep in mind that this name will be displayed on your home grid, so make it something meaningful to you. Tap OK and the collection is created with the name you gave it. The collection displays the first four icons of apps inside it, along with its name underneath (2.27).

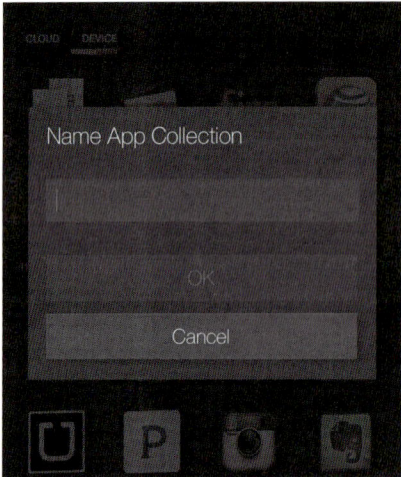

2.26 Enter the name of the collection.

2.27 A new collection, called Fun Stuff, has been created.

▶   **TIP** You can place a collection (or four) in the row of items that appears at the bottom of the Carousel.

Tap the collection to open it and access the apps and media it holds. Each collection has multiple pages, much like the home grid, so feel free to add lots of things to each collection. To add an apps or pinned item to the collection, long tap it and drag it on top of the collection. Release your finger, and the app becomes part of that collection.

To remove items from a collection, tap to open it and then long tap the item you want to remove. Tap Remove from Collection, and it is immediately removed. To delete a collection, remove all the items from it.

Moving a collection works just like moving any other item on the home grid. Long tap the collection until you feel the Fire vibrate, and then drag the collection into its new place. You can move a collection wherever you'd like, with one exception: a collection cannot be placed in another collection.

Pin to Home Grid

You can *pin*, or create a shortcut to, apps to the home grid. Once you've pinned something to the home grid, you can place it in the row displayed on the home screen for quick access if you like.

To pin an app from the Cloud home grid to the Device home grid, long tap the app and then tap Pin to Home Grid in the menu.

Notice that a checkmark isn't displayed on that app in the Cloud tab. This is because when pinning an app from the Cloud home grid to the Device home grid, nothing is downloaded or installed. The app icon appears on the home grid, but the app isn't installed. When you tap that pinned app for the first time, it will download and install itself.

Pinning media to the home grid uses a similar process. You can pin media either from the Carousel or from within the app that plays the media (that is, the Books app for Kindle books or the Video app for a TV show or movie). Long tap the media, and then tap Pin to Home Grid. The media displays on the home grid on your device (2.28).

You can tap this pinned media to begin consuming it, and you can move it just like any other icon on the grid. Once you've decided that you no longer want it pinned, you can unpin it by long tapping it and then and tapping Unpin from Home Grid. This doesn't delete or remove the item from your device; the shortcut is simply deleted from the home grid.

2.28 Books and music can be pinned to the home grid.

Quick Switch

The Carousel displays recently used apps, and the home grid displays all your apps. When you are in an app and want to switch to another (between Maps and email, for example), push the home button, swipe the Carousel until you find the app you're looking for, and tap it to launch it.

That's an awful lot of taps, especially since your Fire phone runs multiple apps at the same time—something called *multitasking*. Quick Switch takes advantage of this multitasking and makes switching between apps a breeze.

Push the home button twice rapidly to bring up Quick Switch no matter where you are in the Fire phone (2.29). Swipe left or right to cycle through the apps that are currently running on your phone.

2.29 Quick Switch allows you to jump from app to app easily.

▶ **NOTE** The Fire doesn't keep running your apps in the background forever, since this would slow down your phone. After a while, apps are closed automatically (without losing data). If an app isn't listed in Quick Switch, launch it from the home grid or the Carousel.

Once you've found the app that you want to switch to, just tap it and you're taken to the app. Press the home button twice and you're back at Quick Switch.

Quick Switch also lets you easily quit apps that you no longer need to run or that aren't responding. Rest your finger on the icon of the app you want to quit in Quick Switch, and flick up. The app is flung off the screen, and no longer running. You can launch it again from the home grid or Carousel.

Keep in mind that just because you can quit apps doesn't mean you should. The Fire's operating system (Fire OS) is designed to manage its resources without user intervention. Feel free to leave apps running—your Fire will handle the rest.

CHAPTER 3

Working with Text

A smartphone is, of course, a phone, but a vast array of its functionality depends on good old-fashioned text. The Fire phone sports a number of ways to both enter and manipulate text. In this chapter, you'll learn how to use the keyboard, enable advanced features, and work with text like a pro.

> ▶ **NOTE** Using your Fire to send text messages to someone (texting) is covered in Chapter 6.

Keyboard

Whenever you encounter a place where text entry is appropriate on your Fire, the keyboard automatically appears (3.1). Each time you tap a letter or number, your Fire vibrates a little to give you some feedback (this is called haptic feedback).

If you need to enter numbers or symbols, just tap the ?123 button. This brings up a keyboard with numbers and commonly used symbols (3.2). For even more symbols, tap the symbol button (~\<) on the number keyboard (3.3). Tap the ABC button to return to the default keyboard.

The Fire keyboard is a software keyboard rather than a hardware keyboard. That's a fancy way of saying there aren't any physical keys on the board; it is all being done on the device itself. This allows the Fire to add and subtract keys depending on what you're doing.

For example, when you're typing a URL in the Silk browser, or an email address in the To field, a .COM button appears on the keyboard. Or when you're asked to type in numbers on certain web forms or apps, a number pad displays instead of the full keyboard.

▶ **TIP** Double tap the spacebar to insert a period at the end of the sentence. The Fire will capitalize the next word after the period.

3.1 Your Fire's keyboard.

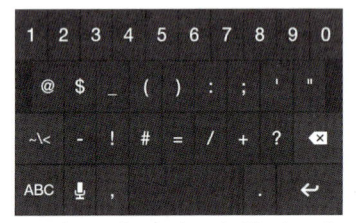

3.2 The number and symbol keyboard.

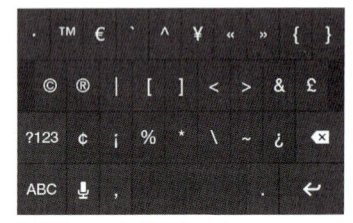

3.3 The symbol keyboard.

Predictive text

As you're typing, the Fire attempts to predict what you're typing and offers some suggestions at the top of the keyboard (3.4). As you type, the suggestions update in real time. The suggestion that is underlined in orange is the word the Fire is pretty sure you meant. If it is correct, simply tap the spacebar and that word will be placed wherever you are typing.

Swipe the suggestion bar to the right to see all the suggestions the Fire is offering. When you find one you want to use, tap it and that word will be inserted.

Once you've typed in a word (or used one of the suggestions), a funny thing happens: more words are suggested without you typing a thing. That's right, based on what you've typed, the Fire tries to figure out what you're going to type. This works best when you're typing common expressions—you can get to "Hello, how are you?" by typing "he" and tapping the suggested words.

Adding things to the dictionary

The Fire phone has an extensive dictionary, but it isn't all-encompassing. Sometimes it won't have any idea what you're trying to type, though it'll still offer suggestions. The unknown word you've typed will be displayed all the way to the left, next to the suggested words (3.5). If this is a word you'll be using often, tap it in the suggestion bar to make the Add to Dictionary option appear (3.6). Tap that button, and you've just taught your Fire phone a word. Good job! The word is added to your personal dictionary, which is unique to your phone, so you are in control of it (unlike the standard dictionary on the phone).

You can add words directly to your personal dictionary. This is handy if you have a long list of words, perhaps jargon or technical phrases that you use often, that you want to add all at once.

3.4 Tap a suggested words to insert it.

3.5 When you type a word the Fire doesn't recognize, it suggests alternatives.

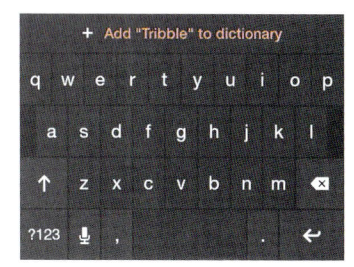

3.6 Tap the alert to add the current word to the dictionary.

CHAPTER 3: WORKING WITH TEXT

To add words directly to the dictionary:

1. Open Quick Settings by flicking your wrist or swiping down from the top of the Fire.
2. Tap the Settings icon.
3. Tap Keyboard > Edit Your Personal Dictionary.
4. Tap Add at the top right, and then type the word you want to add (3.7).

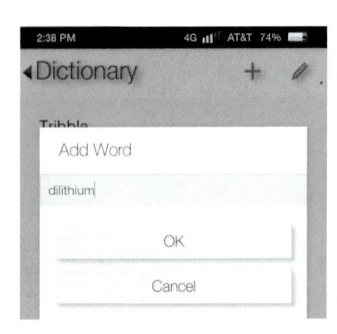

3.7 Type a word to add it to your phone's dictionary.

5. Tap OK, and the word is added.

To edit a word that you've previously added:

1. Open the personal dictionary edit screen by going to Quick Settings > Settings > Edit Your Personal Dictionary.
2. Tap the word you want to edit, and make any changes.
3. Tap OK.

To delete one or more words from your personal dictionary:

1. Open your personal dictionary (see above).
2. Tap Edit at the top of the screen.
3. Tap the check box next to the word or words you want to delete (3.8).

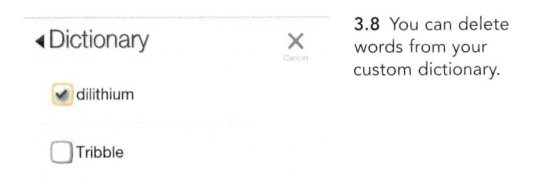

3.8 You can delete words from your custom dictionary.

4. Tap the Delete icon at the bottom of the screen.

Spellcheck

As you type, the Fire automatically checks your spelling, a feature I rely on heavily. Even more helpful, when you type a commonly misspelled word the Fire corrects it automatically. If you don't want to use the correction the Fire has applied, tap the backspace button on the keyboard. If a word is underlined in orange, the Fire is subtly informing you that it believes the word is misspelled (3.9).

3.9 Misspelled words are underlined.

Tapping an orange-underlined word displays suggested replacement words above the keyboard. Swipe to the left to see all the options, and then tap the one you'd like to use as a replacement. If you don't want to use any of them, simply continue typing.

Trace typing

Tapping away at each letter is one way of entering text on your Fire, but it isn't the only way. You can also use something called trace typing.

To trace type, put your finger on the first letter of the word you want to type and, without lifting your finger from the screen, slide your finger to the next letter and the next and so on until you've slid to every letter in the word. An orange line traces your path on the keyboard to let you know that your Fire is following along (3.10).

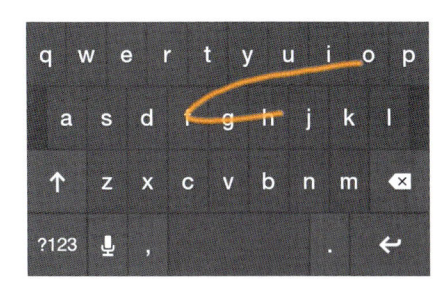

3.10 Swipe your finger along the keyboard to type words.

CHAPTER 3: WORKING WITH TEXT   31

The Fire will pop the word it thinks you meant into the text field and offer you some suggestions, as it always does. If your Fire can't figure out the word you meant, a message asking you to try again will display at the top of the keyboard (3.11).

3.11 If your Fire can't determine what word you've swiped, it asks you to try again.

Speech to text

There is one more method of text input on your Fire: speech to text.

On the keyboard, you'll see a microphone icon. Tap it and a microphone icon with a circle of "fireflies" will appear superimposed over the keyboard. Say what you want to type, including punctuation, and then tap Done. The text—or at least what the Fire thought you said—will be inserted. You can also just stop talking, and after a few moments the Fire will sense that you've stopped talking and automatically insert the text for you.

> **NOTE** Speech to text is intended for dictating short messages, so it stops listening after a few moments and inserts text automatically. If you need to dictate a long message or a large amount of text, you'll have to tap the microphone icon several times.

Transcribing doesn't happen in real time. When the fireflies turn into a square on the screen, that signifies the transcription is in progress. What's happening here? The transcription doesn't actually happen on your phone. Instead, the Fire takes the recording of what you said and sends it to Amazon's servers to be transcribed. Once the servers do the transcription, they send the data back and the text appears on your screen. Sending your voice data to remote servers allows Amazon to apply much more computing power to the process than is available on your phone. While in theory this speeds up transcription, it does have two downsides:

- The transcription doesn't happen live, so you have to say a sentence or two at a time.

- You have to have an active network connection (cellular or Wi-Fi). If you try to use speech to text without a network connection, you'll get an error message (3.12).

3.12 Speech is unavailable when you don't have a network connection.

▶ **NOTE** You can do more things with your voice on your Fire. I cover those functions, and the potential privacy concerns, in Chapter 14.

Advanced keyboard

Out of the box, the Fire keyboard gets the job done, but with a simple setting switch you'll make it much more powerful. The advanced keyboard allows you to use alternate characters quickly and adds a few other keyboards.

To enable to advanced keyboard:

1. Open Quick Settings, and tap the Settings icon.
2. Tap the Keyboard section to expand it, if needed, and then tap Manage Advanced Keyboard Features.
3. Toggle Advanced Keyboard from off (the default) to on (which is way more awesome).

Alternate characters

Notice that each key now has a label in white and then a smaller label in gray. When you tap a key, the white character will be typed. But when you long tap a key with a light gray label in the upper-left corner, a menu appears with alternate characters that you can type (3.13). These characters include numbers, accented letters, and symbols. Tap one, and it will be inserted in the text field and you'll return to the keyboard.

3.13 Long tap a key to get alternate characters with the advanced keyboard.

CHAPTER 3: WORKING WITH TEXT

Other keyboards

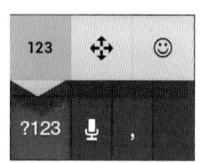

Even with the advanced keyboard turned on, when you tap the ?123 button it brings you to the symbol/number keyboard. The magic happens when you long tap it (3.14). Now you have access to three alternate keyboards.

3.14 Long tap to switch keyboards.

The number pad keyboard is there for you when you need to type a long number or a simple math problem (it won't do the math for you, of course) (3.15). Long tap the ?123 and then tap the 123 button (slightly confusing to read, but it is clear in the interface), and you'll be taken to the number pad. Tap the ABC button to return to the regular keyboard.

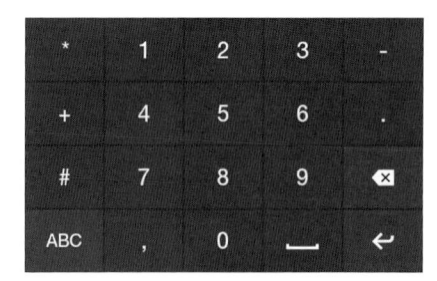

3.15 The number pad.

The text selection keyboard is dedicated to all things related to text. Long tap the ?123 button and then tap the button with the four arrows on it. This takes you to the text selection keyboard. Here's what each of the buttons do:

- Tapping this button when you have text selected cuts it from the text field and allows you to paste that text elsewhere.
- Tap this button to copy the selected text for pasting elsewhere.
- Tap this button to paste previously copied or cut text.
-  Tapping this button inserts a few spaces or jumps to the next field on a form.
- Tapping the middle button here engages text selection mode (and makes the button light gray so you know the mode is engaged). Using the arrows, you can select text. Once you have the text selected, you can use the other buttons to perform copy, paste, or cut operations.
- Tap this button to select all the text onscreen.
- The Delete key.

- ▭ The spacebar. Double tap it to insert a period.
- ↵ Tap here to start a new line.
- ⬅ Tap this button to return to the regular alphanumeric keyboard.

The final alternate keyboard is all about emoticons. Sometimes words cannot express your feelings—that's where emoticons come in.

These little faces composed of punctuation marks have taken the world by storm, and they're easy to insert when you've enabled the advanced keyboard. Tap the happy face button (as shown in 3.14), and the emoticons menu will appear (3.16). Tap an emoticon to insert it into the text field. If you decide against an emoticon (why would you do that?), tap the X to close the menu and get back to the keyboard.

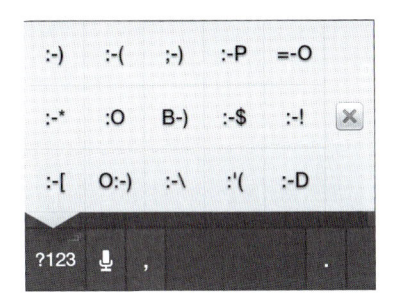

3.16 Emoticons ;).

Selecting Text

You're no doubt familiar with the concepts of copying, pasting, and cutting. They mean the same thing on your Fire, but since there's no mouse button to click, you get to those options by tapping. This is also how you select the text that you want to copy, cut, or paste.

To select text, you can certainly use the text selection keyboard, as detailed in the previous section, but that's best used when you're doing a lot of copying and pasting. For one word or sentence, it is far easier to use your finger to select the words and then copy/paste them:

1. Tap a word you want to select.

2. The Fire vibrates, and the word you tapped is selected (and highlighted in blue). Two handles appear on either side of the word, along with menu options (3.17).

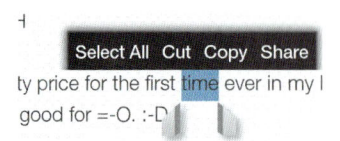

3.17 Tap a word to select it, and use the handles to select more text.

3. If you're just interested in selecting one word, then you're done. If you want to select more than one word, like a sentence or a whole paragraph, drag one of the handles on the selected word. A magnifier will appear so you can more accurately select what you're after (3.18). Lift your finger off the display when you've selected all the text that you're interested in.

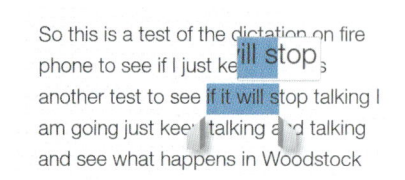

3.18 The magnifier gives you a better look at what you're selecting.

Once you have the text selected, turn your attention to the menu that appears above your selection. This menu displays different options depending on the context in which you've selected the text:

- When you select non-editable text—say, text on a website or in a text message—you'll have two options: Select All and Copy. Tap Select All to highlight all the text on the current screen. Tap Copy to copy the selected text to the clipboard. You'll have to switch to a text field to paste the copied text.

- When you select editable text—text you've typed into the Notes app or a text field in Silk, for example—you'll have a few more options. Our old friends Select All and Copy are here, and Cut and Paste join the party. Paste replaces the selected text with whatever text was previously copied to the clipboard, and Cut removes the selected text but saves it to the clipboard so you can paste it elsewhere.

Depending on the app, there may be an additional item on the menu: Share. Tapping Share brings up the Share panel, which lists all the apps you can "share" that text with (3.19). For example, on my Fire phone I have several apps that can take the selected text and perform an action:

3.19 The Share panel lists apps that will take the text you've selected.

- Email creates a new email message with the text.
- Add to Evernote creates a new note in Evernote (a notes app that syncs across devices).

Since you'll have different apps on your phone, you'll see different options on the Share panel, but it functions in the same way. Tap the app that you would like to send the text to, and it will perform an app-appropriate action.

Inserting the cursor

Sometimes in an email you'll want to correct a typo or add a word between two other words. You can do this on the Fire, though only on editable text.

When you're in a text field (or app) with text that you can edit, tap the place where you'd like to place the cursor, and it appears with a handle at the bottom (3.20). When you type, the text will be entered at that point and the cursor's handle will disappear. If you tap the Delete button, anything immediately before the cursor will be deleted.

3.20 The cursor has a handle so you can move it with precision.

CHAPTER 3: WORKING WITH TEXT   37

To move the cursor elsewhere, either tap in a different location or tap the cursor handle and drag it. As you drag, the magnifier appears, allowing you to see text that might otherwise be obscured by your finger and letting you place the cursor with greater accuracy.

CHAPTER 4

Email

Email is a part of our culture, and it is expected that you can read and respond to emails no matter where you are. With your Fire you can do this (assuming you have a network connection) and even more.

Once you've read this chapter, you'll be the master of email on your Fire, though your Fire can't automatically clear out your inbox for you—yet.

The Fire supports several kinds of email accounts:

- POP accounts download your email messages from the mail server and store them on your Fire. Once they've been downloaded, they can be deleted from the server.

- IMAP is the setting I recommend to people. Instead of downloading email to your Fire and deleting it from the server, with an IMAP account the mail stays on the server but is readable on your device. Lose your Fire? Your email is still on your mail server, so it is safe and sound.

- Exchange. Many corporations use Exchange for email, contacts, and calendaring. The Fire supports all of those uses.

▶ **NOTE** Contact and calendar syncing is also supported for Yahoo, Gmail, and Outlook email addresses.

Setting Up an Email Account

If a major email provider supplies your email account, chances are that it will be very easy to set up on your Fire. If, however, you run your own email server or get email from a smaller provider (perhaps your web host or a super nerdy friend), then you'll have to know a little bit more before you can start checking your email.

Let's start with the easiest setup first, the large providers:

1. Open the Quick Settings panel by swiping down from the top of the Fire, and tap the Settings icon.
2. Tap the My Accounts section.
3. In the expanded My Accounts section, tap Manage Email Accounts.
4. Tap Add Account (4.1).

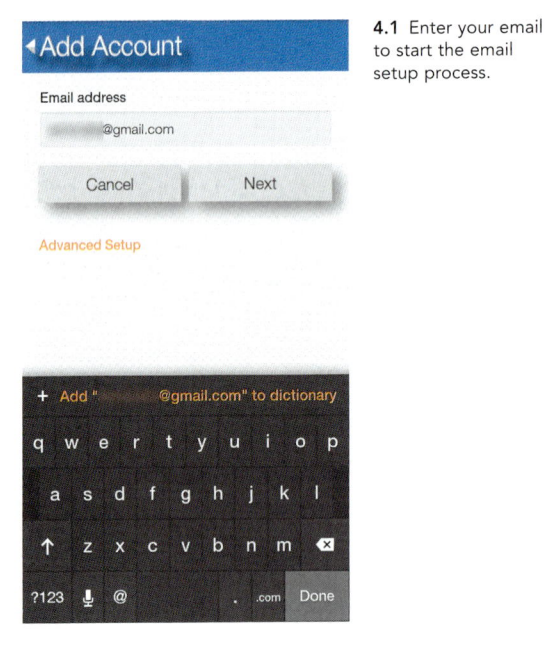

4.1 Enter your email to start the email setup process.

5. Your Fire automatically configures the email settings based on the email address you entered. If you've entered a Gmail account, the Fire sends you to Gmail's authorization page (4.2). Enter your Gmail account password and tap Enter. If you have two-factor authentication

enabled on your account (if you don't know what this means you don't have it enabled), enter your one-time code and tap Enter. Google then displays the list of permissions you're giving your Fire by adding this account to the phone. If you don't want to add the account, tap Cancel; otherwise, tap Authorize and the account is added.

For all other email account types, a password field appears (4.3). Enter your password, but keep in mind that by default the password is displayed as you're typing. If you want the password to be displayed as a series of asterisks to foil snooping eyes, tap the Hide Password box.

Tap Next when you've finished entering your password.

That's it! Your email account is set up. Your Fire is busy downloading email (and your contacts and calendar, if appropriate) in the background. This process can take a few moments. You can go right to this account's inbox or add another email account if you like (4.4).

▶ **TIP** If this is the first email account you're adding to your Fire, you can get to the same screen by sliding out the right panel on the home screen and tapping "Tap to add an account for email and calendar updates" or by tapping the Mail app's icon.

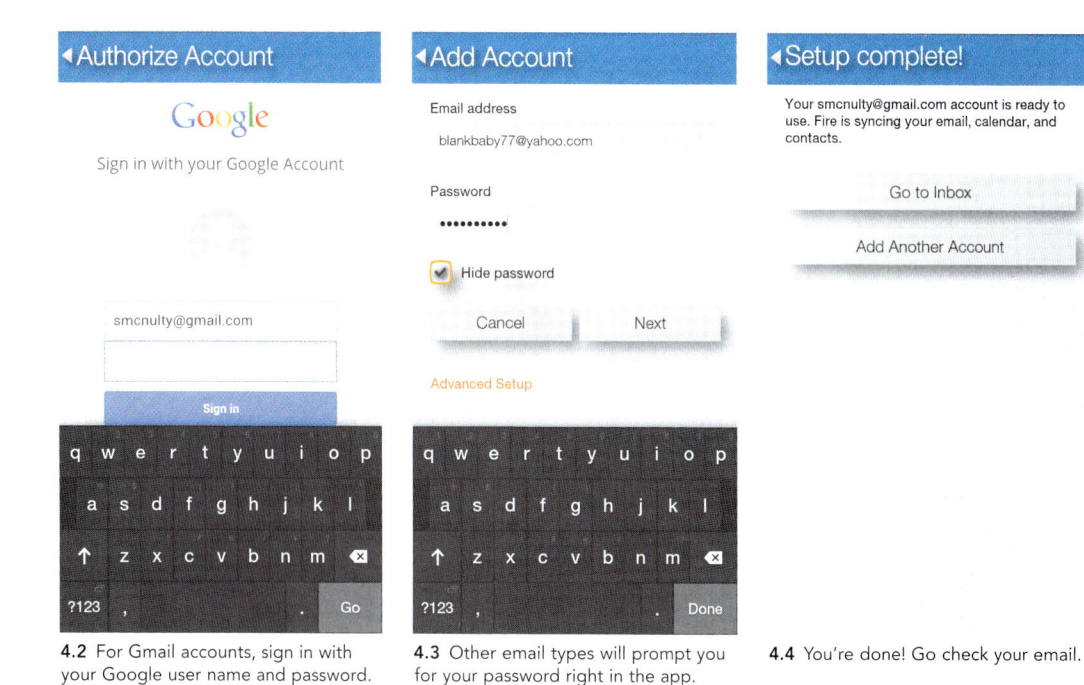

4.2 For Gmail accounts, sign in with your Google user name and password.

4.3 Other email types will prompt you for your password right in the app.

4.4 You're done! Go check your email.

CHAPTER 4: EMAIL   41

Exchange Accounts

Many corporations give people Exchange accounts for email, calendaring, and contacts. Exchange is Microsoft's email server product, and as you might imagine, it was designed to be run in big companies. Why should you care about this? Well, if you have an Exchange account, your Exchange administrator (that's the team or person who runs the Exchange servers for your company) can set security policies. These policies can be applied to your Exchange account and might force you to change some settings on any device that you connect with your Exchange account (your Fire phone, for example).

For example, my Exchange admins decided that if I connect my Exchange account to a mobile device, that device must have a password on the lock screen and must be encrypted.

By default, neither of these things is true of a Fire phone. When I added my Exchange account, the Fire warned me that the Exchange server wanted to change some settings on my phone and asked if I was OK with that **(4.5)**. I tapped OK, but if I didn't want a lock code or encryption on my device I would have had to have tapped Cancel. I wouldn't be able to check my Exchange mail, calendar, or contacts, but I wouldn't have to change my phone's settings either.

Organizations impose these requirements for a simple reason: phones are easy to lose. If you lose a phone with super secret corporate documents on it, wouldn't it be cool to prevent ne'er-do-wells from just swiping up and seeing everything on the phone?

Encryption, lock codes, and other security-related topics are covered in Chapter 16.

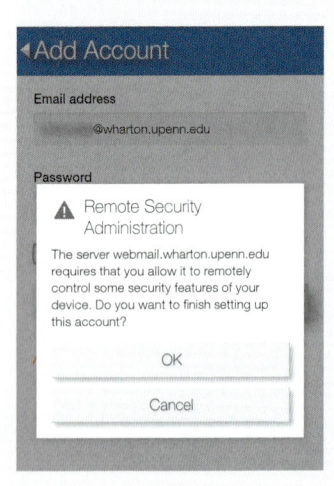

4.5 Exchange accounts can require heightened security settings.

If your email account's settings aren't automatically detected by the Fire, you'll need to know some information about your account to successfully set it up:

- Whether your account is POP or IMAP (Exchange accounts are almost always auto-detected)
- The mail and SMTP server addresses

If you don't know this information, consult your email provider's documentation. This is basic information about your account, so it shouldn't be difficult to track down.

Once you've gathered that information, the process is very similar to the one for setting up a Gmail account:

1. Swipe down to open Quick Settings, go to Settings > My Accounts, and tap Manage Email Accounts.

2. Enter your email address and tap Next. The Fire asks you to enter your password so it can try to auto-configure your account.

3. The Fire will fail to detect your account's settings, so it will ask you to configure them manually (4.6). Along the top of the screen, you'll see the three types of email accounts supported by the Fire. Since you've already gathered some information about your email account, you should know if it is POP, IMAP, or Exchange.

   For POP accounts, tap POP3 (4.7).

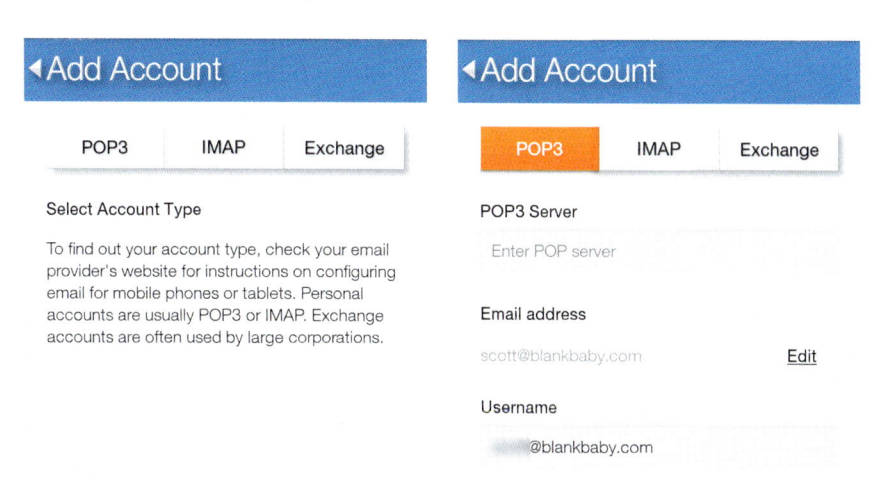

4.6 Manually adding an account allows for email accounts from smaller providers.

4.7 Enter the details of your email account.

CHAPTER 4: EMAIL   43

Enter your POP and SMTP server details (the email address and password you previously entered are automatically filled in here). By default, your Fire will download email but won't delete it from the server. Tap Delete Email from Server to change this behavior to "When I delete from Inbox." This will make it so that when you use your Fire to delete an email in this account, it will be deleted from the email server as well.

The IMAP settings are nearly identical except for one setting: Default Folder. Enter the default folder for your IMAP account if you have one (consult your email provider's documentation to find out if you need to set this; most people do not).

Tap Exchange to manually set up an Exchange account, which might be needed if your organization is running an older version of Exchange. You'll need the Exchange server address, your user name, and your password. You might also need the domain for your email.

No matter which account type you are setting up, you can also look at the security settings by tapping Security Settings and Ports. The defaults should work 99 percent of the time. The documentation for your email account will tell you very clearly if you need to change the security settings or port numbers, and this is where you can do that.

4. Tap Next, and your email account is successfully set up. If contacts or calendar sync is supported, that will also start working for this newly added account.

▶ **TIP** You can always access these settings when adding a new email account by tapping Advanced Settings after entering an email address in the setup screen.

Deleting Accounts

Deleting an email account from your Fire is simple:

1. Open the Quick Settings panel, and tap the Settings icon (or tap the Settings icon in the Carousel if it is there).

44   THE AMAZON FIRE PHONE

2. Tap My Accounts, and then tap Manage Email Accounts.
3. All your accounts are listed under the Accounts heading. Tap the one you want to delete.
4. Scroll down to the Remove Account section, and tap Delete Account from Device (4.8).

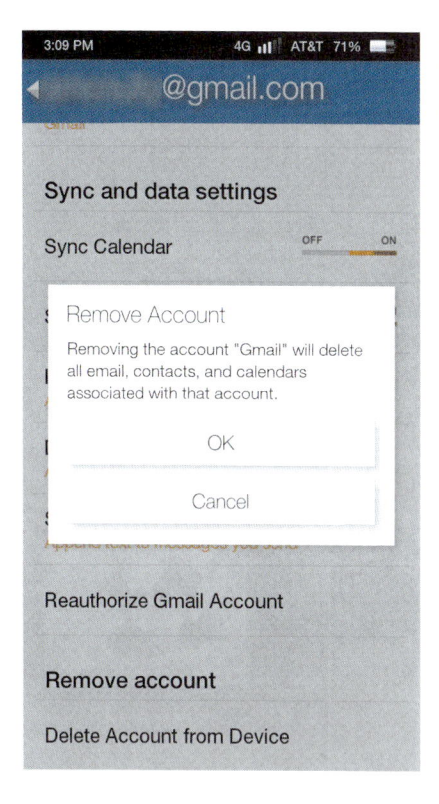

4.8 Deleting an account is easy. Just the data on your phone is deleted.

5. Tap OK.

The account has been removed, though all of the information associated with that account is still available on your provider's servers (unless this is a POP account that's set to delete email from the server).

Account Settings

While we're in the account settings for one of your email accounts, let's take a look around and see what you can set here (4.9).

Display

The email address of the currently selected account is displayed at the very top of the Email Settings screen.

The Your Name field is the name that is displayed when an email from you appears in someone's inbox. The Description field shows what this email account is named in the Fire interface. Tap either one to edit its text.

Default Account is the account that the Fire will use by default when creating new email messages. Tap here to change the default to any currently linked email account (4.10).

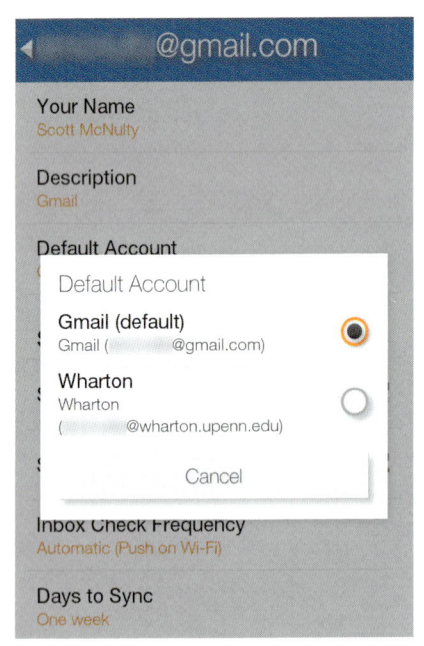

4.9 The email settings also list all the accounts on your phone.

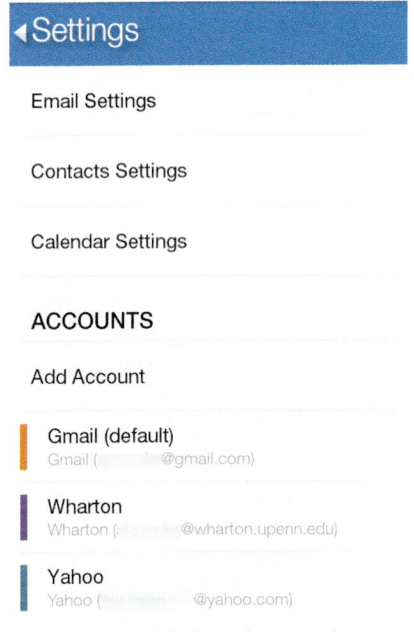

4.10 Make your default email account the one you use most often.

Sync and data settings

Setting up an email account wouldn't be of much use if it didn't actually sync your email to your phone. But whenever you get an email on your Fire, that process uses a little data (or a lot of it, depending on the message). This doesn't matter much when you're on Wi-Fi, since you can generally use as much bandwidth as you like without consequences, but when you're on a cell network you can use only as much data as your plan allows before you enter "overage charges" territory. No one likes that.

The Sync and Data Settings section controls what gets synced with your Fire on a per account basis and how often this syncing happens.

The Sync Calendar and Sync Contacts toggles turn on and off syncing of calendar and contacts for supported accounts. Tap the toggle to change it from its current position (on to off and vice versa).

Inbox Check Frequency tells your Fire phone how often to poll this account for new email (4.11):

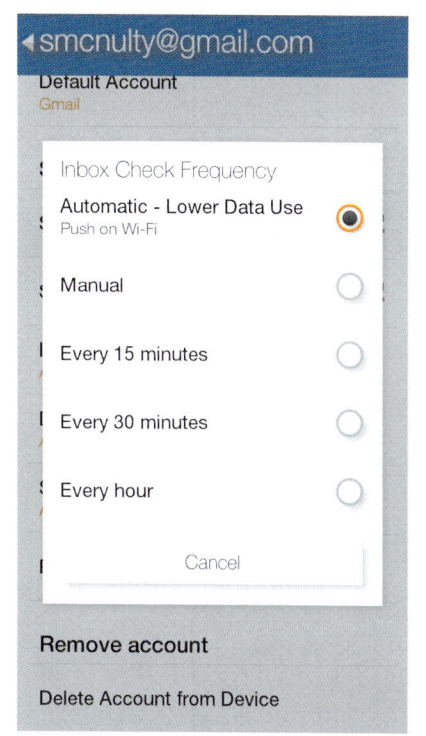

4.11 Set the frequency your inbox will update.

CHAPTER 4: EMAIL 47

- Automatic: Also called "Push," this option is available for certain supported email accounts (Gmail and Exchange, to name two). When an email is delivered to your inbox on the remote server, the server "pushes" the new message to the Fire. The Fire doesn't need to check at all; the server just sends the new message and it appears on your phone instantly.

    For some accounts, like Gmail, when this option is selected it will work only when your Fire is connected to a Wi-Fi network. When connected to the cell network, the Fire will check infrequently in order to use less data.

- Manual: New mail is checked for only when you initiate the check yourself. The Fire won't automatically check for email. This is a great setting when you're on vacation, or if you're very concerned about data usage, because you are in total control over when you will get new email.

- The remaining options allow you to set a schedule for when your Fire will check for email. Select Every 15 minutes, Every 30 minutes, or Every Hour, and your Fire will check for email on that schedule. You can also manually check in case you're anxiously awaiting a particular email.

Days to Sync tells your Fire phone how much of your email you want to have available on the device (4.12). You can select Automatic and leave it up to the Fire, select a particular time period (one day to one month), or just select All to have all the email in that account available on your device. You can change this setting at any time, so I recommend starting with one week and then adding more time if you find yourself looking for older email. When changing this setting to a longer period, you should be connected to a Wi-Fi network; faster downloads speeds will make the sync process quicker.

You can append a signature to every new email you send from your Fire. In fact, the Fire thoughtfully appends the signature "Sent from my Fire" to new emails by default. If you'd rather your Fire didn't do this, or you want to change the signature to say something else, tap Signature and edit the text (4.13). Enter new text, or delete all the text, and tap OK.

Signatures are per account, so your work account can have a signature with your work phone number and fax number, while your personal account could have a whimsical or inspirational saying.

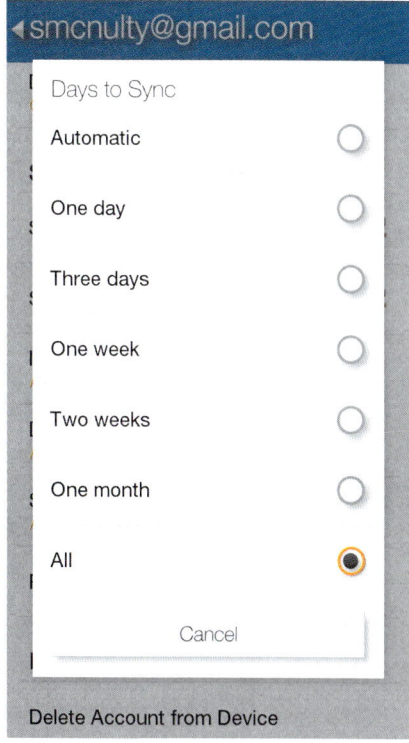

4.12 Days to Sync controls how much email is downloaded to your phone.

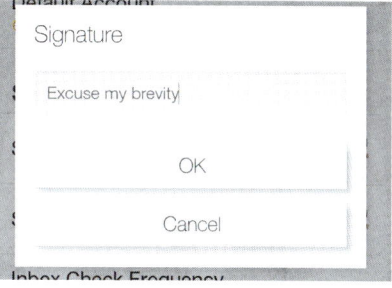

4.13 This signature will be inserted into all email sent from your Fire.

Server Settings are found at the bottom of the settings screen for each account type (other than Gmail). Tap this to change server settings, login information, and more.

At the bottom of the settings screen for Gmail accounts, you'll find a Reauthorize Gmail Account option. This allows you to get back in to your Google account should you change the password.

Checking and Reading Your Email

Now that your accounts are set up, it's time to actually check your email. Launching email is as simple as tapping the Mail icon at the bottom of the home screen.

CHAPTER 4: EMAIL 49

You can see a few recent messages in the preview beneath the Mail icon in the Carousel (4.14). Tap one of those previews to go directly to that message in the Mail app.

Tap the triangle under the icon on the Carousel to set what is included in the preview (4.15):

- VIP: You can designate certain contacts as VIPs, which allows you to highlight messages from them. More about VIPs in Chapter 5.

- Combined Inbox: Combined Inbox is what I have my Carousel preview set to because I have two email accounts on my Fire: my work email and my personal account. This view shows me email from both inboxes together, sorted chronologically. The emails themselves are still in the individual accounts; this view just makes it easy to see all your email across your inboxes at once.

- A particular account: Display only emails from one account.

4.14 Email is previewed on the Carousel.

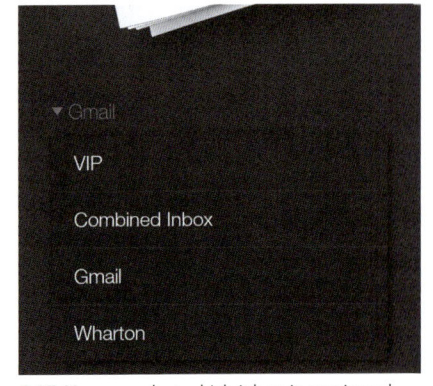

4.15 You can select which inbox is previewed.

Tapping the Mail app icon brings you to your inbox. The upper-left corner tells you what account and folder you're in. In **4.16** we're looking at the combined inbox. Unread messages are in bold, and each message has a colored bar on its left edge. Each account gets its own color so you can quickly see which account this message was sent to. The sender's name is displayed above the subject and a brief preview of the message.

Swipe up to scroll through the messages, and swipe down to return to the top of the message list. If you pull down (swipe down and then release) while you're at the top of the message list, the Mail app checks for new messages.

The mail app's left panel allows you to jump to different email accounts or directly to folders within those accounts (**4.17**). Just swipe or tilt to the right to bring the panel up, and tap the account or folder you're interested in.

Tapping Search reveals a search field at the top of whatever view you're currently looking at (**4.18**). You can search the From, To, or Subject fields by tapping one, entering your search term, and then tapping the Done key. The search results appear. Tap an email to see the whole thing.

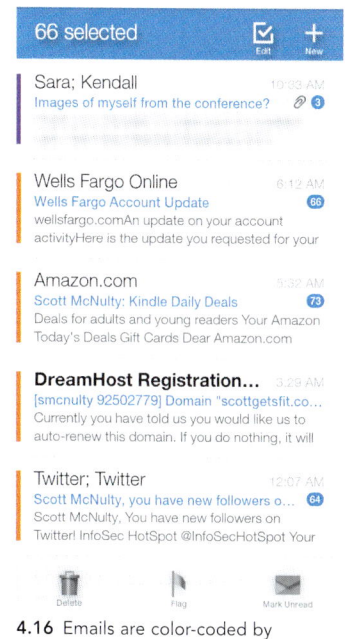

4.16 Emails are color-coded by account in your inbox.

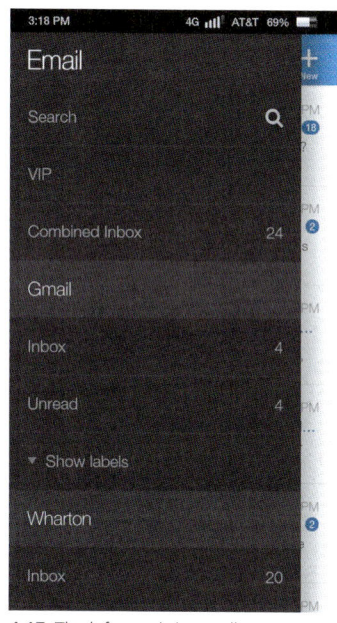

4.17 The left panel shows all your accounts and folders, as well as special folders like VIPs and the combined inbox.

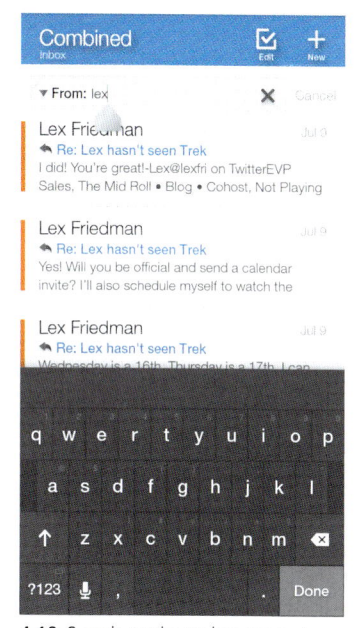

4.18 Search results update as you type.

CHAPTER 4: EMAIL   51

The right panel in the inbox or folder view is pretty neat (4.19). It lists all the attachments in that folder or inbox. Swipe up and down to go through the list. If the attachment has been downloaded, just tap it to open it, but if it still needs to be downloaded to your Fire (if it has this icon on it ⤓, the attachment needs to be downloaded), the first tap will start the download process. When the attachment has been download, tap it again to open it.

When you're in an email, the upper-left corner shows the icon of the person who sent you the email (4.20). This is either the picture assigned to this person in your contacts or their initials. Right next to that are their name and the date you received the email. Below that are the To field and a downward arrow. Tap the arrow and you'll see the time and date that the email was sent.

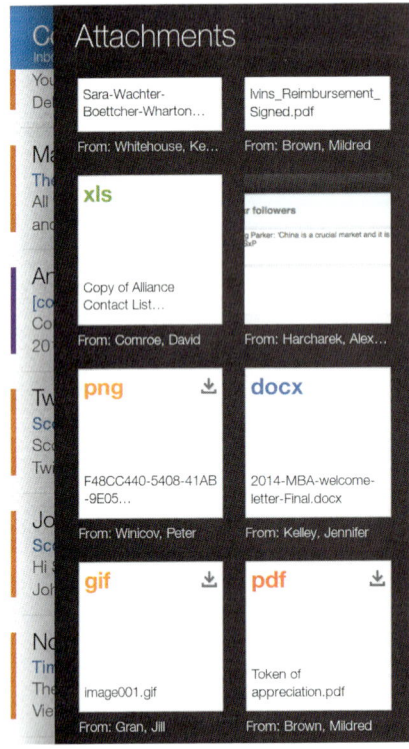

4.19 The right panel displays all the files attached to emails in your inbox.

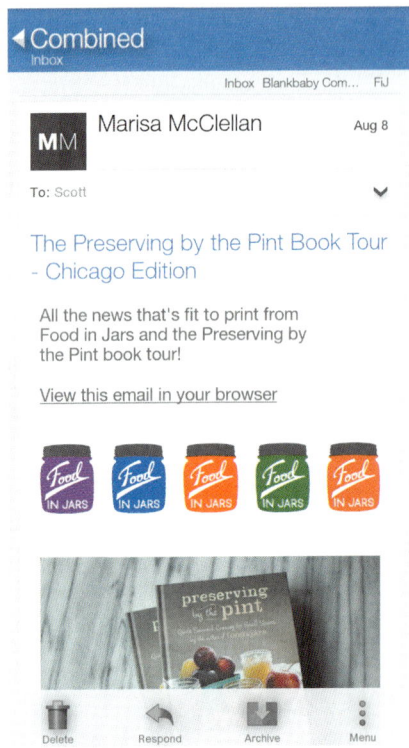

4.20 An email.

The subject of the email is displayed in blue, and below that is the body of the message. In some cases you'll see a Show Complete Message button. That's because that message was rather large, so the Fire didn't download the whole thing. Tapping that button displays the entire message.

Along the bottom of every email message you'll see four buttons:

- Delete: Deletes the email.

- Respond: Tap this button and you get three options: Reply (send a reply to the sender), Reply All (send a reply to the sender and anyone in the CC field), and Forward (send this email to someone else so they can have a copy).

- Archive or Delete: For Gmail accounts, you'll see an Archive button, which will move the message out of your inbox and into your archive so you can search for it later. All other email accounts have a Delete button to trash the message.

- Menu: Tapping this reveals a number of options. **Move** brings up a list of the folders in that email account (4.21). Tap one of the folders and the message is moved into it. **Flag** puts a little flag next to the email so you can reference it later. **Mark Unread** changes the state of the message from read to unread (it will be displayed in bold in the Inbox view). Label is another one of those Gmail-only options; "labels" are Gmail's folders. Tap Label and a list of the available labels is shown (4.22). You can apply as many labels as you like to one email message by tapping the check mark next to each label. Uncheck labels by tapping once more. Tap Apply and the labels are shown in small letters at the top of the email. Tap **New Message** to create a new, blank email.

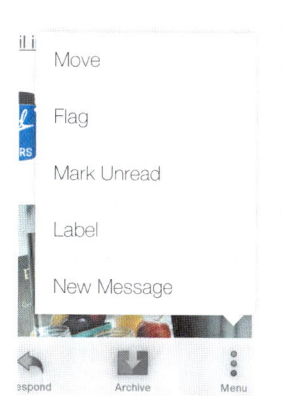

4.21 The Menu button gives you access to some useful functions.

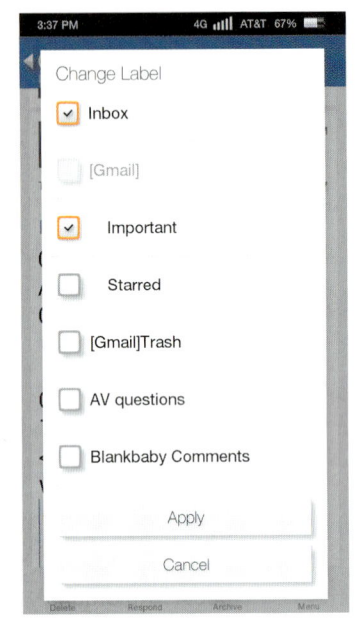

4.22 Labels in Gmail are supported. Tap to add a label.

4.23 While you're in an email, the right panel displays al the email you've received from the sender.

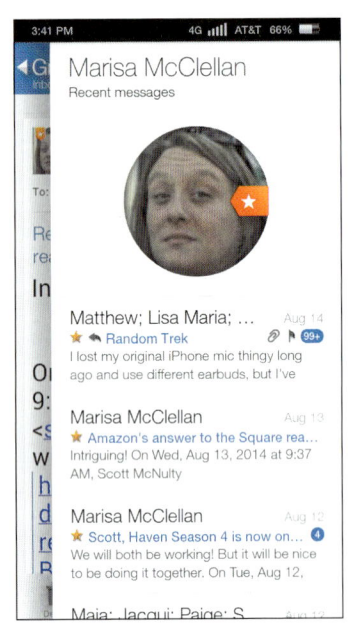

4.24 Marisa is now a VIP in my contacts, as she should be.

While you're reading an individual email, the right panel (swipe from the right edge to the left, or tilt the right edge toward you) shows you recent messages from the same email address **(4.23)**. Scroll through emails by swiping up and down, and open one by tapping it. Tap an avatar (the circle with either your contact's picture or their initials) to mark that person as a VIP, denoted by an orange star **(4.24)** (more about VIPs in Chapter 5). Swipe or tilt to the right to dismiss the panel and return to the message.

You can quickly go to the next or previous email by swiping from the center of the message you're reading. Swiping to the right goes to newer emails (based on the date and time the email was delivered), and swiping left brings you to older emails.

To get back to your inbox, you can either tap the name of the inbox at the upper-left corner of the screen or bring out the left panel and tap whatever inbox or folder you want to hop into.

Composing Email

When you're in an inbox view, tap the New button in the upper-right corner to create a new, blank email (4.25). The From field will automatically be set to your default email account. Tap it to expand the CC/BCC and From fields. You can then tap From and select a different email address to send from (4.26).

Once you've settled on an email address to send from, type an address into the To field (adding addresses to the CC and BCC fields works the same way). As you type, your Fire will suggest email addresses of your contacts (4.27). Tap the contact that you want to send to, if they're listed. If not, continue typing the email address.

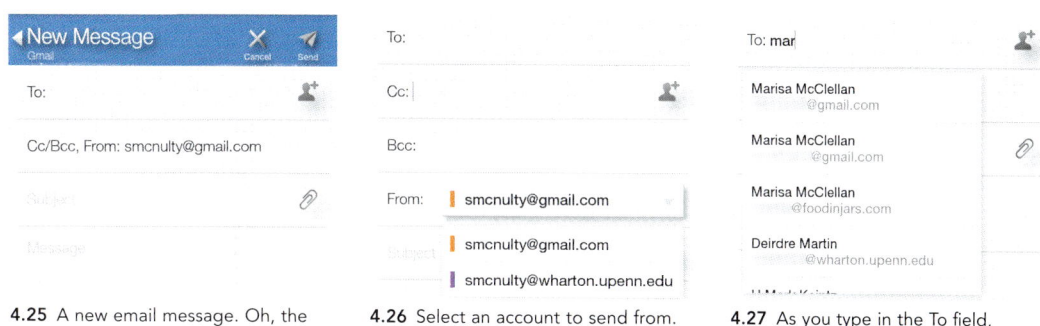

4.25 A new email message. Oh, the possibilities.

4.26 Select an account to send from.

4.27 As you type in the To field, suggestions from your contacts appear.

Alternatively, if you know the person you want to email is in your contacts, you can tap the Contacts icon. This brings up your list of contacts. Scroll through them, or enter a name in the search field at the top (4.28). Tap the person you want to send to.

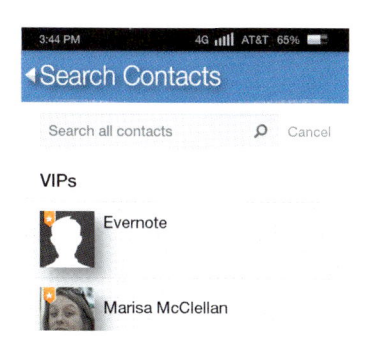

4.28 You can search your contacts directly.

You can also quickly add VIPs and recent contacts to an email message by opening the right panel (4.29). Your VIPs and recent contacts are listed. Tap any or all of them to add them to the message.

If you're emailing someone in your contacts and they have more than one email address listed, how do you make certain you're sending the email to the right address? Tap their name in the To/Cc/Bcc field, and a menu will drop down that includes all the email addresses for that contact (4.30). The email address to which this email is addressed has a checkmark next to it. To send to one of the other email addresses listed, just tap the address and the checkmark moves to that address.

To remove addresses from any of the recipient fields, tap the address and tap the Delete key on the keyboard.

▶ **TIP** If you want to send a message to multiple email addresses that aren't listed in the right panel, you'll have to add each email address separately.

Tap the Subject field to add a subject to your email. At the end of the Subject field is a paper clip icon that allows you to attach files to your email. Tap the paperclip to bring up the Attachments menu, which has the following options (4.31):

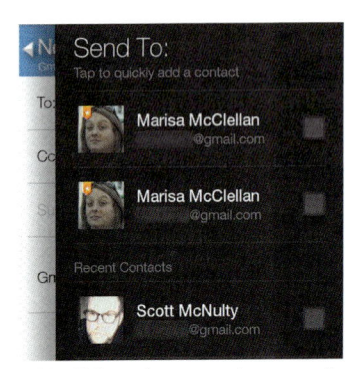

4.29 While you're composing an email, the right panel displays VIPs and recent contacts.

4.30 For contacts with multiple emails, tap the To field to select another email address.

4.31 You can attach photos, videos, or files to email.

56 THE AMAZON FIRE PHONE

- Attach a Photo: Tapping this brings up the pictures on your Fire's camera roll (4.32). Tap the one that you want to attach to your email, and you'll see a large preview with two icons below it. Tap the X to pick a different photo; tap the checkmark to select this one. When you select a photo, it is added to the email in the attachment section under the message field (4.33).
- Attach a Video: The same as the above, only with videos.
- Attach a File: When you tap this option, your Fire will ask you to select an app from which you'd like to attach files. Tap the app, and then select the file you want to attach.
- Capture a Photo: Tapping this launches the Camera app so you can take a new picture and attach it to this email. Point the camera at whatever you want to snap a picture of, tap the shutter icon, and then tap the checkmark to attach it to your email (tapping the X cancels the process and returns you to your email; tapping the Camera icon discards the image you just took but allows you to take a new one). Taking pictures with your Fire is covered in Chapter 10.
- Capture a Video: Same as the above, only with videos. Be careful of file size when sending videos via email—they can be large, and many email servers will reject email messages with overly large attachments.

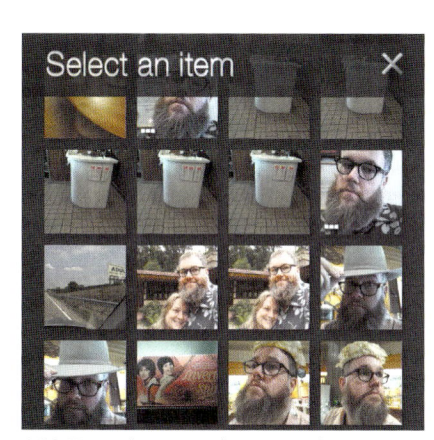

4.32 Tap a photo (or video) to attach it.

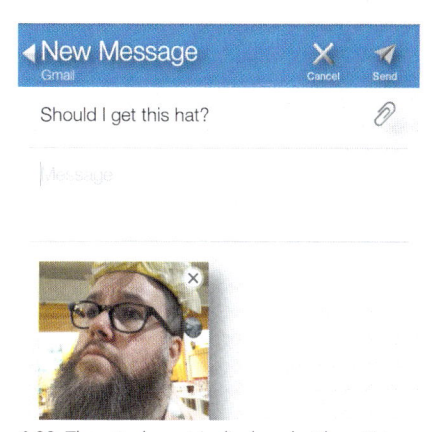

4.33 The attachment is displayed with an X icon. Tap to delete.

An email message can have multiple attachments. Keep adding attachments until you're done; they will all be displayed in the attachment area. Each attachment will also sport an X icon in the upper-right corner of its icon. Tap the X to remove that attachment from the email.

You've got attachments, a recipient, and a subject, so now all you need to do is type a message (though that is optional). Tap the message's body field and type (or dictate) your message. There's no limit to the amount of text you can include. You can even copy and paste text from other sources into your email.

Tap the Cancel button in the upper-right corner to discard a message. Tap the send icon to email the message. An animation indicates that the message has been sent, and you're taken back to your inbox.

▶ **TIP** You can compose emails even when your Fire is in Airplane mode or otherwise disconnected from networking. When you send messages without a data connection, they are queued and sent as soon as the Fire connects to a network.

Managing Email

Once you're back in a folder or inbox, you can manage your email. Swipe to the left on any of the messages in the list to reveal two icons: Move and Delete (**4.34**). Tapping Move brings up your folder list, into which you can move the message. Tapping Delete immediately deletes the message.

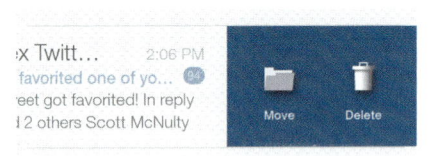

4.34 Swiping an email reveals these buttons.

▶ **TIP** Swiping an email in the Carousel will reveal a Delete button, so you can delete emails without going into the Mail app.

If you want to move or delete a large number of emails at once (or a number greater than one for that matter), tap the Edit button in the

58   THE AMAZON FIRE PHONE

upper-right corner of the folder you're in. All the emails will slide to the right, and checkboxes will appear to their left (4.35). Select the emails you want to edit by tapping them. As soon as one email is selected, the edit action buttons appear at the bottom of the display. When you're in your combined inbox, you'll have the actions Delete, Flag, and Mark Unread. If you're in a Gmail account, you'll see Delete, Archive, Move, and a Menu button that contains Flag, Mark Read/Unread, and Label.

Delete, Mark Read/Unread, Move, and Label have been covered. Flagging a message applies a little flag to the message so that you know it is notable for some reason (4.36). These flags do show up in other email clients, so you could, for example, flag emails that you want to respond to when you get back to your computer.

▶ **NOTE** When you have only flagged emails selected, the Flag button turns into an Unflag button.

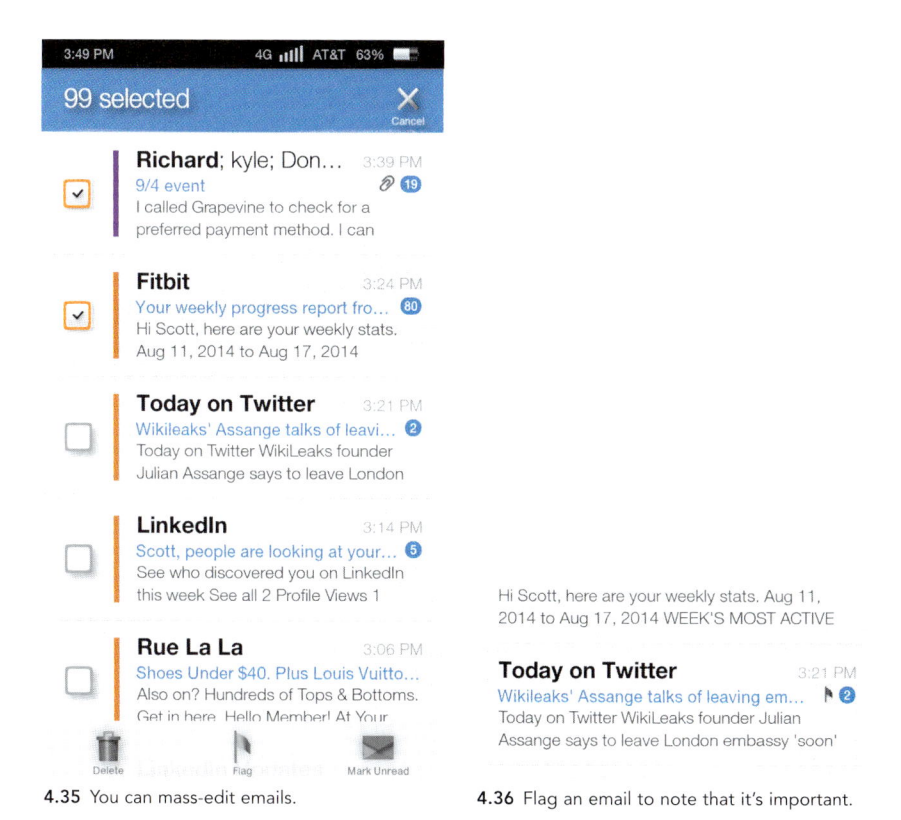

4.35 You can mass-edit emails.

4.36 Flag an email to note that it's important.

CHAPTER 4: EMAIL 59

Archive, for Gmail accounts, removes the email from your inbox but doesn't delete it. The email is in your Archive folder, so you can find it later.

Tap Cancel to close out of editing mode.

You can also enter editing mode by long tapping an email in the list. This does the same thing as tapping the Edit button, with an added bonus: the email you long tapped is automatically selected, saving you a tap.

Customizing Your Mail

Slide out the left panel while you're in your email, and swipe down until you see Settings in the More section. Tap Email Settings, and you'll see a few settings that you can play around with (4.37).

By default, the text in the Mail app is kind of small. Tap Default Message Text Size to choose a new size from five options (I usually go with Medium) (4.38). The font size is changed immediately, though it only impacts the text size of email messages, not the rest of the Fire's interface.

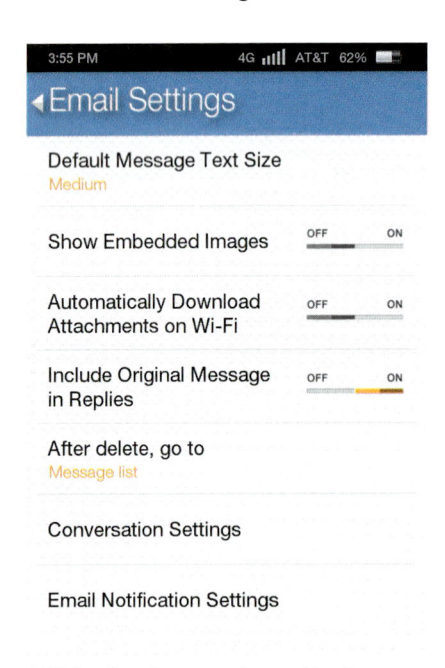

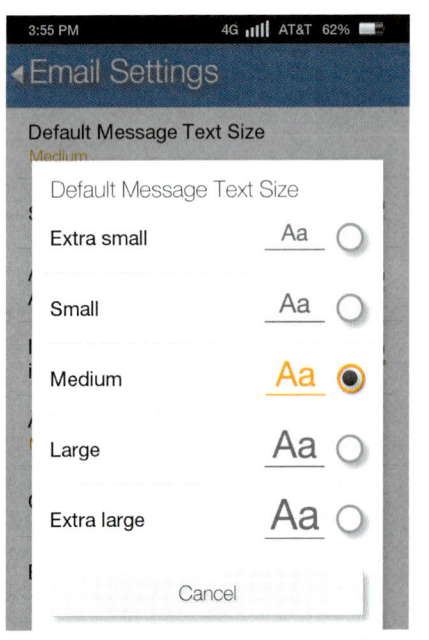

4.37 Email settings control everything from text size to the downloading of attachments.

4.38 The five text sizes available in the Mail app.

Next up are three options that can be toggled on or off:

- Show Embedded Images: Leave this on if you want images in emails to be shown automatically. When this setting is off, a Show Images button appears in emails with images (4.39). Tap it to show images in just that email. The button then changes to "Always show images from this sender" (4.40). Tapping that will automatically show images in all messages from that person in the future.

4.39 Tap Show Images to download and display the images in this email.

4.40 Show images once, and you can tell your Fire to always display images from this sender.

- Automatically Download Attachments on Wi-Fi: Off by default, this setting will download attachments only when your Fire is connected to a Wi-Fi network. This will save you a tap (usually you have to tap to download the attachment and then tap again to open it), but keep in mind that your Fire has limited storage capacity. If you often get emails with large attachments, you should probably leave this off.

- Include Original Message in Replies: You'll notice when you reply to an email on your Fire, the original email's content is copied into the new email and your response is on top. If you don't like this, toggle this off and the new email will just have your response. I find that it is helpful to include the original email in responses because it maintains some context, but it is a stylistic choice.

The next three settings affect how you interact with your mail.

After you delete a message, you are taken back the message list of whatever folder you were in. Tap "After delete, go to" for two other options: Newer Message and Older Message. Tap one, or tap Cancel to keep the setting as it is.

Tap Conversation Settings to see the options for grouping conversations (4.41). Your Fire can group, or display together, conversation threads in the message lists. If you toggle this on, your Fire will group any emails with the same subject line together in the message list and display a blue number that represents the number of emails in that conversation (4.42). You can move and delete the conversations as a group.

When you tap a conversation, the most recent message is displayed with excerpts of the previous messages displayed above (4.43). Tap any excerpt to see the full contents of that message.

Now you'll notice that only the emails that I received are included in the conversation display. If you turn on Conversations Include Sent and Drafts Folders, then your responses (and responses you are drafting) will be included in the conversation display.

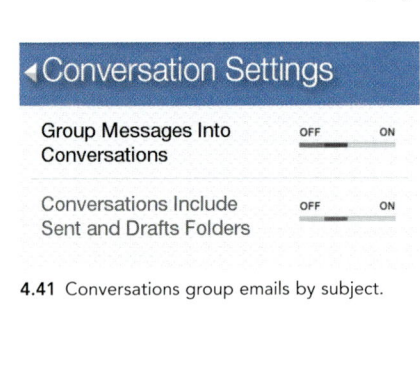

4.41 Conversations group emails by subject.

4.42 Conversations display the number of emails in the conversation to the right of the subject line.

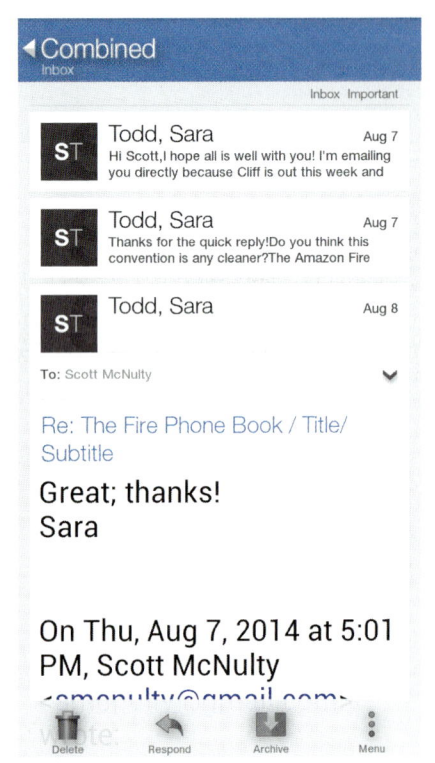

4.43 Reading a conversation.

62   THE AMAZON FIRE PHONE

Swipe up or tap the left arrow at the top of the conversation settings to return to the Email Settings screen, and tap Email Notification Settings.

Your Fire has several ways it can notify you of new email. By default, it will use notifications to show you a preview of new email (4.44). Swipe down from the top of the screen to see your notifications; they're listed under the Quick Settings.

In 4.44, I have seven new messages in my Wharton account. Notice the three lines in the middle of the notification. If I swipe down on those lines I get a quick preview of the messages (4.45). Tap a notification to be taken to the message list so you can read, respond to, or delete the message.

If you don't want these notifications, toggle them off at the top of the email notification screen.

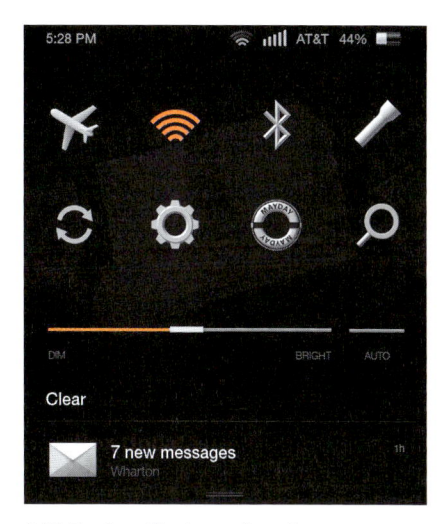

4.44 Email notifications tell you how many new messages you have.

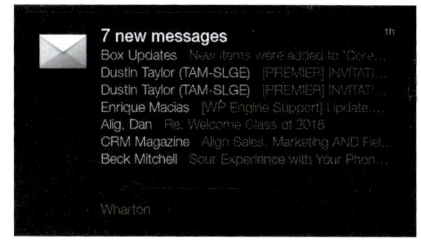

4.45 Expand to see more information about the email.

You can also toggle on banner notifications here. Banners slide down from the top of the display no matter what you're doing and show you the subject of the new email and who sent it (4.46). They slide back up after a couple seconds, but if you're a quick tapper you can tap a banner notification to be taken to the email. Don't worry if you miss a banner when you're trying to tap them, because they are available in the notifications area of the top panel (swipe down from the top of the Fire) until you dismiss them (by swiping either to the left or right along the notification).

CHAPTER 4: EMAIL   63

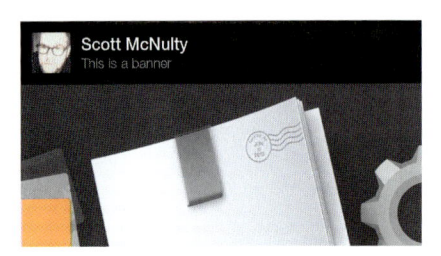

4.46 Banner notifications slide down from the top of the screen.

If notifications and banners aren't enough, you can also have your Fire make a sound and vibrate every time you get a new message. Keep in mind that these notification settings are shared across all the email accounts on your phone, so you can't set a sound to play when you get email in your work account and your phone to vibrate for your personal email. It is all or nothing.

Tap No Sound (which is the default) in the Sound section to choose which sound is played when you get new email. The Fire comes with a number of different alert sounds, and as you tap one to assign it the sound plays. Swipe up and down to scroll through the list to find that perfect sound. Once you've found the one, tap the checkmark at the top right of the screen to return to the notification settings. The Sound section will now display the name of the alert you picked (you can come back here at any time and change this sound to another or to change back to no sound).

Right under the Sound section is the Vibrate section. When this setting is on, your phone will vibrate every time you get a new email message.

When all the notifications are turned on, this is what will happen when you get a new email in any account:

- An alert sound plays and your phone vibrates.
- A banner slides down from the top with a preview of the email.
- A notification appears.
- A badge on the Mail icon displays the number of new messages.
- The Carousel preview shows you the latest messages in whatever view you have it set to.

You'll have no excuse for missing an email with all these settings on!

CHAPTER 5

Contacts

The Contacts app is your address book. It lists the contact details for friends, colleagues, and businesses, and it makes it easy to manage all those details in one place. As you use it to generate emails and text messages to your friends, it will even display your recently used contacts in the Carousel so you can get to them with one tap **(5.1)**.

This chapter covers everything you need to know: setting up a profile, adding and managing contacts, assigning custom alerts and photos, VIP contacts, and importing and exporting contacts from your Fire.

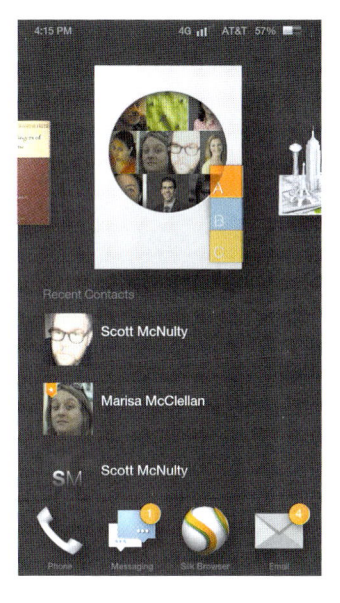

5.1 The Contacts icon on the Carousel uses your contact pictures.

Getting Started

You're probably already managing your contacts either through your email account or on your current phone.

If you're using your email account to keep your contacts (such as in Yahoo, Gmail, or Exchange), when you add that account to your Fire those contacts will sync to the Contacts app on the phone. There's no need to import or export anything.

To make sure your contacts are syncing with your Fire, open the email account settings (Quick Settings > Settings > My Accounts > Manage Email Accounts) and, under the Accounts heading, tap the account that you want to sync contacts with. Make sure that Sync Contacts (under the Sync and Data settings) is set to On.

In the Contacts app, the left panel (swipe from the left edge to the right or tilt the left edge toward you) displays some navigational options (5.2). Tap one to see the contacts from that source. To go back to seeing all contacts, tap All Contacts.

▶ **TIP** You can sync contacts from multiple accounts to your Fire phone.

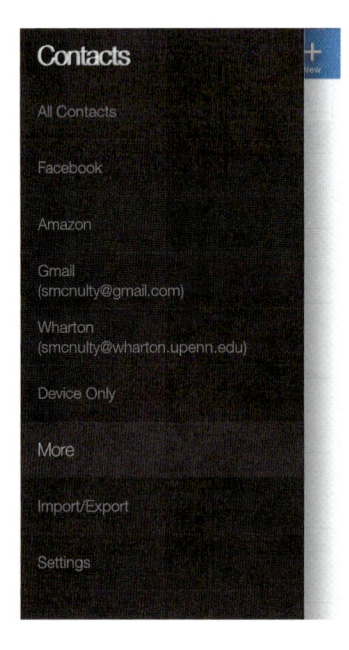

5.2 The left panel lets you move from one account's contacts to another, or look at all your contacts.

Transferring Contacts from Your Old Phone to the Fire

If you've been keeping your contacts on an old smartphone and not syncing them to another service, you'll need to transfer them to your Fire phone. You can do this on Android phones and iPhones by downloading the AT&T app on both your Fire phone and your old phone and following the instructions on Amazon's site:

Android instructions: http://amzn.to/1qisUW7

iPhone instructions: http://amzn.to/1kN3YJH

If you have a non-smartphone with contacts you'd like to transfer to your Fire, stop by an AT&T store. Make sure to bring both your old phone and the Fire. The folks there should be able to transfer your contacts for you.

You can also import and export contacts from a phone's SIM card by opening the left panel in the Contacts app and tapping Import/Export.

The easiest way to import contacts into your Fire is to export them from the old phone to a VCF file (an industry-standard contacts format). Once you have the VCF file, transfer it to your Fire by emailing it to yourself and downloading the attachment to the phone, or you can transfer it to your Fire via USB (see Chapter 10 for more information about this method).

Once you have the VCF file on your phone, import by doing the following:

1. Open the Contacts app.
2. Reveal the left panel, and tap Import/Export.
3. Tap Import From Phone Storage (5.3).

5.3 Importing and exporting your contacts is easy.

The phone looks at all the local storage until it finds a VCF. Once the VCF is found, the import process begins (5.4). You're notified when the import is done, and then the contacts are listed in the Contacts app.

5.4 The import process starts with a notification letting you know it is happening.

Setting Up a Profile and Contacts

To launch the Contacts app, go to the home grid and tap the Contacts icon, which displays some of your contacts' pictures, if you have any (5.5).

If this is the first time you're launching the Contacts app, the first thing listed is "Set up my profile." This is your personal contact in the Contacts app.

Tap it to see all the fields you can fill in (5.6). Your profile is structured just like every other contact, so we'll use it as an example for creating contacts in general.

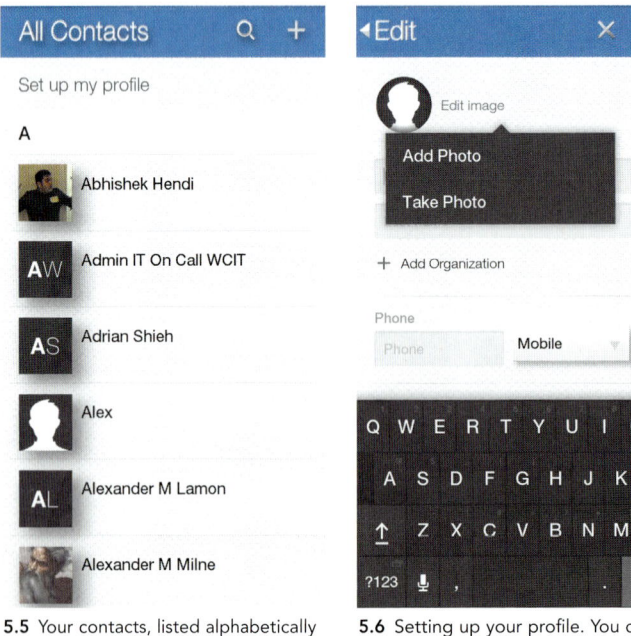

5.5 Your contacts, listed alphabetically by first name.

5.6 Setting up your profile. You can include a picture of yourself.

68   THE AMAZON FIRE PHONE

To add a picture to your profile, tap Edit Image (next to the default avatar) and either add an existing photo from the photos on your Fire or take a photo with the Fire. Tap Take Photo to launch the camera app.

Once your profile (or a contact) has a picture associated with it, the options in the Edit Image menu change to Remove Photo, Select New Photo, and Take New Photo.

Enter your name by tapping in the fields and typing. Tap the arrow next to the first field to show a few more name fields (5.7). Tap Add Organization to list your work organization (or any other organization you're involved with) and title.

The Phone and Email fields have drop-down menus next to them that allow you to label the location associated with that number or email address (home, mobile, work, and so on) (5.8).

You can add your own type of label to the Phone menu by tapping Custom, typing a name for your custom label, and tapping OK (5.9). Now that label will show up in the list whenever you add a phone number to a contact on your phone.

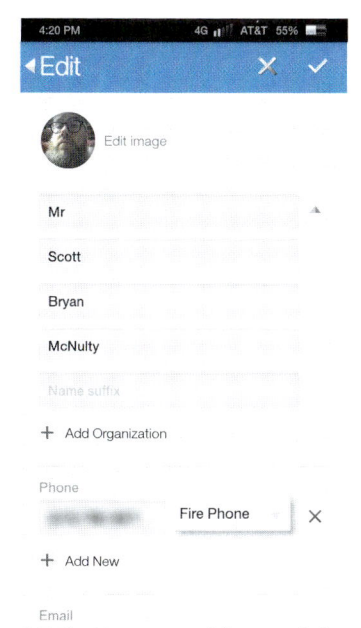

5.7 The Name option fully expanded.

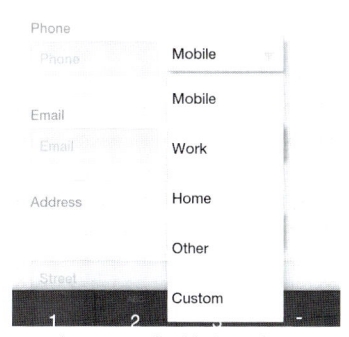

5.8 The options for labeling phone numbers, including custom settings.

5.9 The custom label can say whatever you like.

CHAPTER 5: CONTACTS

As soon as you enter text into either the Phone or Email field, a few new options appear: Tap the X to clear the value you just entered; tap Add New to add an additional phone or email address. (I have four email addresses and three phone numbers. I may be too connected.)

The Address section works much the same way. There are only three labels available: Home, Work, and Other. Once you start typing in the fields you can add another set of address information by tapping the Add New button under the address you're entering.

Under the address fields is a button labeled Add More Fields. Tap that to add even more information to your profile (and contacts): Nickname, Website, Events (which only covers birthdays and anniversaries at the moment), and Notes.

Once you're finished filling in the information, tap the checkmark in the upper-right corner of the screen to save (or tap the X to discard).

You're taken to your completed profile, which looks just like a contact (only better since you're such a fine-looking individual) (5.10). Your picture, if set, is displayed front and center, with all the information you entered below it.

In your profile, as in any contact, you can take some actions by tapping the relevant piece of information. Each phone number is listed twice. Tap the top phone number to call it, and the one below it to send a text. Tap the email address to start a new email.

Long tap either piece of information to copy it to the Clipboard for pasting elsewhere.

Tap the Address field to send it to the Maps app, which will make it easy for you to find directions (more on using Maps in Chapter 10).

Along the bottom of your profile, and every contact, you'll see the same three buttons: Delete 🗑, Edit ✏️, and Share 🔗. Edit and Delete work as you would expect them to. Tapping Share brings up the Share panel, with a number of ways to share this contact with someone (5.11). Keep in mind that your list of sharing options will depend on the apps you have installed on your Fire.

In contacts other than your profile you'll see a couple of additional things at the bottom of the contact. First, you'll see the source of the contact: either the account that the contact is synced with or Device, which means that this contact is only on your Fire. Then you'll see the menu icon ⋮. Tap this and you'll see two choices: Join and Split. They're discussed in the "Joining and splitting contacts" section later in this chapter.

5.10 My full complete profile. Aren't I a handsome devil?

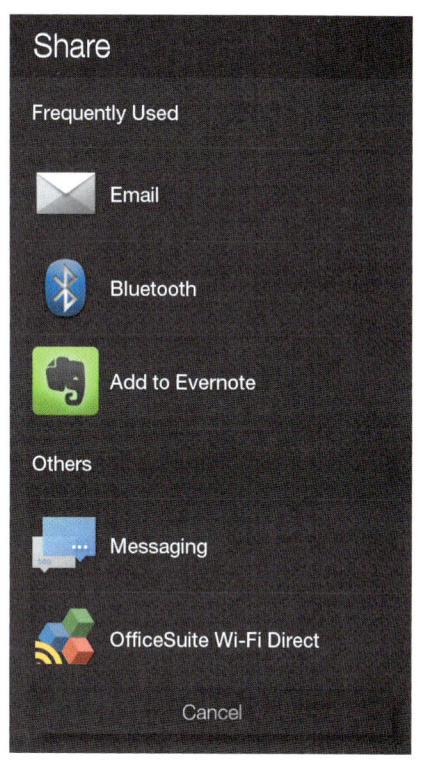

5.11 Sharing a contact brings up the Share panel.

Tap the arrow at the top of the screen to return to the contacts list, and you'll see that there is now a Me entry at the top. That's your profile.

▶ **TIP** Use Peek to see the last time you've contacted your contacts (5.12).

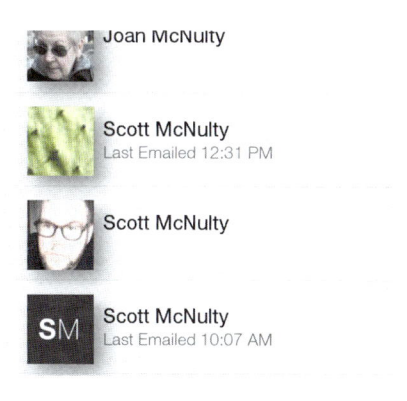

5.12 Peek shows when you've last emailed or called a contact. I need to call my mother!

CHAPTER 5: CONTACTS 71

To add another contact directly on your device, tap the + button in the upper-right corner of the contacts list. The contact entry screen works and looks exactly like the Profile screen, with one exception: customized ringtones (5.13).

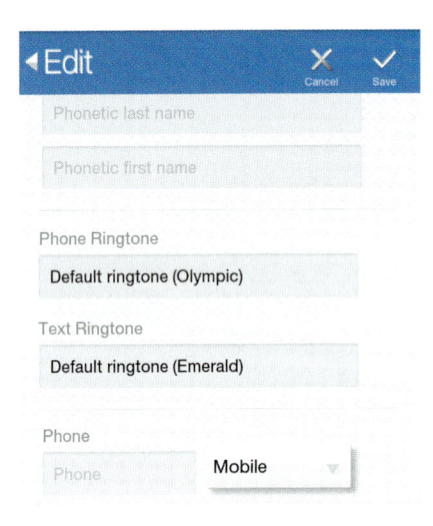

5.13 Assigning a custom ringtone for calls and texts.

You can set custom ringtones for calls and texts on a per-contact basis. Tap Default to see a list of ringtones or alerts. Tap the Ringtone field to pick a new one from a list. Tap to select it, and it will play. You can also choose None for both phone calls and text messages so that your phone won't make a sound whenever that person calls or texts. Tap the checkmark in the upper-right corner to set the phone or text ringtone. Your choice will be displayed in the contact, and you can change it at any time by editing the contact and selecting a new ringtone.

Navigating contacts

In the contacts list, swipe up and down to scroll through your contacts. Your contacts will have a photo (if set) or a gray box with their initials next to their names. By default, contacts are listed alphabetically by first name. Tap a contact to open it.

If you are looking for a particular person but don't want to scroll, tap the magnifying glass in the upper-right corner to bring up the search field (5.14). As you type, your contacts will be filtered until only those that match the search value are displayed. You can even continue the search on the server that hosts your contacts (if the contact you're looking for isn't on your device) by tapping the Search on Server button.

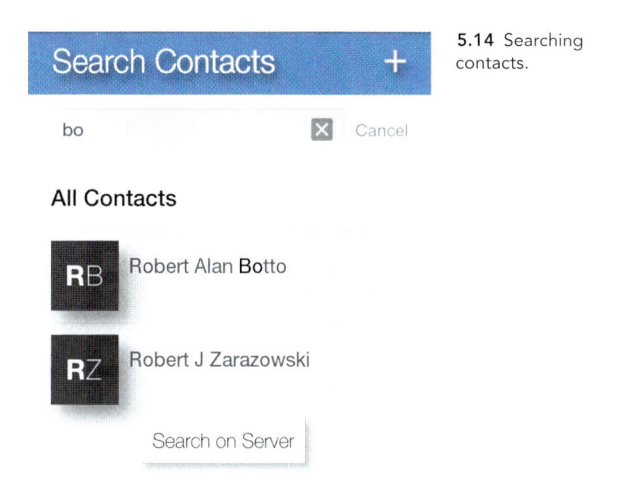

5.14 Searching contacts.

Tap a contact in the results to open it, and then tap the phone number or email address to contact them.

VIPs

You can set any contact as a VIP. This means that you're especially interested in messages, calls, and texts from these people. VIPs might include your children, your boss, and your significant other.

To make someone a VIP:

1. Open the Contacts app, and find their contact.
2. Tap it to open it.
3. Tap the star on their avatar.

    It turns orange, and they are now a VIP (5.15).

The benefits for being a VIP on a Fire phone include a special VIP-only mail view and quick access to VIP contact information using the right panel in the Contacts app and phone app (5.16).

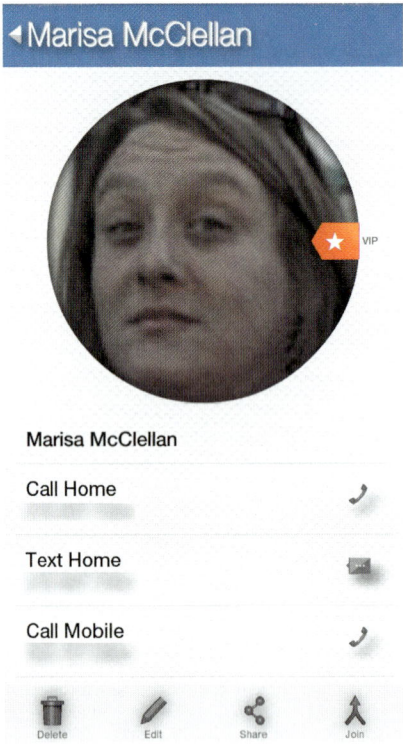

5.15 Marisa is a VIP.

5.16 The right panel lists your VIPs.

Contacts quick view

When you're in the contacts list you can quickly access information about a contact without actually going into the contact. This makes it very easy to email or call someone without having to use too many taps. Here's how to do it:

1. Find the contact you're looking for, and long tap their avatar (their picture or the initials next to their name).

2. The Contacts quick view appears (5.17). From here you can make someone a VIP by tapping the star, or you can call, text, or email them by tapping the relevant button. This view shows you only the information that you can use to initiate contact. You won't see their address, nickname, or other fields. View the entire contact by tapping View Contact.

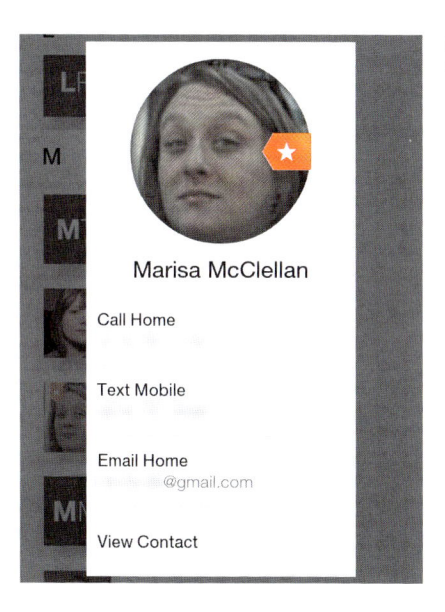

5.17 The quick view screen.

Marisa McClellan

Call Home

Text Mobile

Email Home
███████@gmail.com

View Contact

Joining and splitting contacts

If you sync more than one source of contacts, you might end up with multiple contacts for the same person, each with different pieces of information. You can join these disparate contacts into one contact easily:

1. In the contacts list, long tap the contact you'd like to join.
2. A menu appears with two options: Join and Share. Tap Join **(5.18)**.

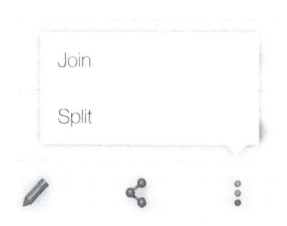

5.18 The Split and Join commands.

A list of suggested contacts to join with the selected contact appears, along with a list of all your contacts **(5.19)**.

3. Tap the contact you want to join with the original contact.

To split a joined contact back into its component contacts:

1. Tap the contact you want to split.
2. Tap the menu icon, and then tap Split.
3. Tap OK, and the contact is split into two (5.20).

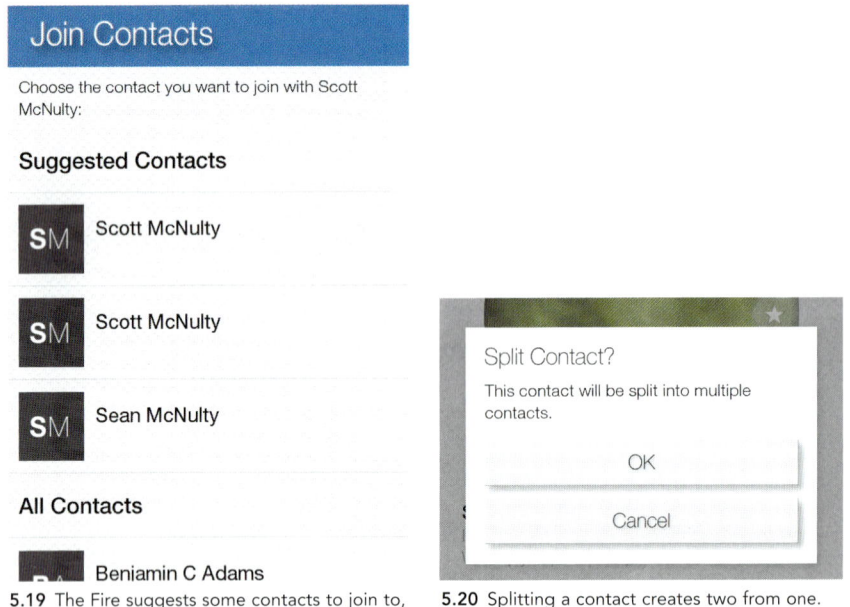

5.19 The Fire suggests some contacts to join to, and allows you to pick any contact.

5.20 Splitting a contact creates two from one.

Settings

There are a few interesting Contacts settings that you should be aware of. You can get to settings in a couple of ways:

- Open the left panel in the Contacts app, and tap Settings.
- Go to Quick Settings > Settings > My Accounts > Manage My Email Accounts > Contact settings.

Either way, you get to the same screen of settings (5.21).

5.21 Contacts settings.

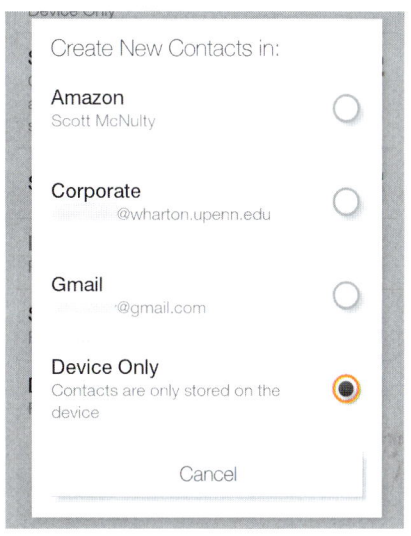

5.22 Contacts created on your phone are saved in this list.

First, you need to set where you would like new contacts to be created. By default, contacts you create on your Fire are stored only on your Fire. They won't be synced to any accounts, so you won't be able to access them anywhere other than on your Fire.

To change this, tap Create New Contacts In (**5.22**). Tap the account you'd like new contacts to be created in, and the panel will close. Notice that the top entry is simply called Amazon. By default, your contacts sync to your Amazon account for backup purposes.

If you don't like the idea of your contacts syncing to Amazon, you can disable it. Toggle Synchronize Contacts to off. This will only stop syncing with your Amazon account; other accounts will continue to sync contacts.

Disabling syncing only stops the synchronization; it doesn't delete what has already been synced. To really be sure that none of your contacts are on Amazon's servers, tap Delete Amazon Contacts From Cloud. This will clear out any contacts that are on Amazon's servers. You can always repopulate them by enabling the sync process.

Sync Facebook Contacts syncs your Facebook contacts to your Fire. Tap On, and you'll be prompted to set your Facebook account on your Fire. If your Facebook account is connected to your Fire, just tap On and your Facebook contacts will appear on your phone.

The last two settings affect how contacts are sorted (Sort Order) and how they are displayed (Display Order). By default, your contacts are sorted alphabetically by first name. To change this, tap Sort order and select Last, First. The same option exists for the display name. By default, names are displayed as "Scott McNulty"; that's First, Last. If you'd rather see "McNulty, Scott," tap Last, First.

CHAPTER 6

Messaging

Sending and receiving text messages (SMS and MMS) happens in the Messaging app on your Fire **(6.1)**. This chapter covers how to check your messages, send some of your own, and quickly text photos and videos. Sending to more than one person and changing the Messaging app settings is touched upon as well.

Open the Messaging app by tapping it in the home grid or the Carousel. You'll see the message list **(6.2)**. This list will be empty if you don't have any text messages, but as you can see, I have two conversations because I'm very popular (let's not worry that one is from my wife and the other is from myself to myself). Note that each message in the list represents a threaded text message conversation. The most recent message in the conversation is previewed, but there could be several messages in the conversation that aren't visible on the message list.

The top menu tells you that you're in the Messaging app and displays the number of unread messages (if any). Next to that are the search icon and the new message icon. Before we tap either of those, let's take a closer look at the message list.

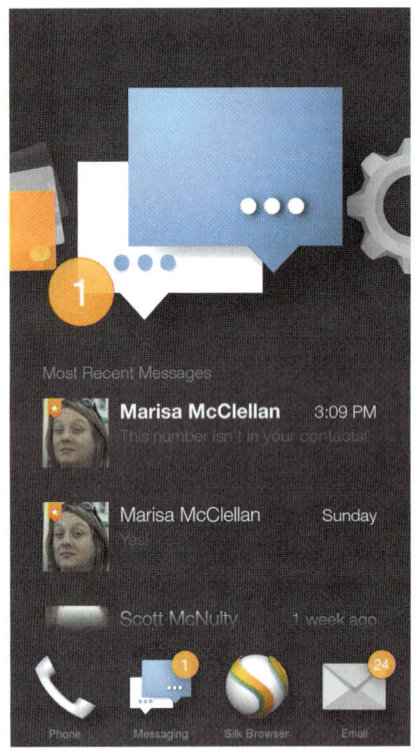

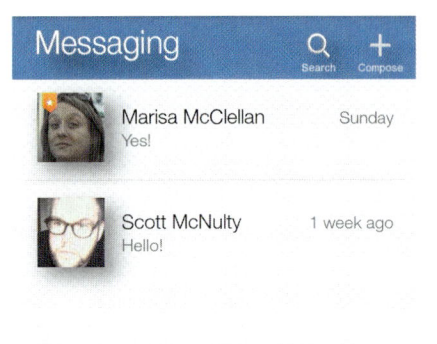

6.1 The Messaging app in the Carousel displays recent messages.

6.2 The message list shows a preview of the most recent message in the conversation.

Each conversation is listed in chronological order, with the newest on top and unread messages bolded. The time or date the message was received displays to the right. If the number messaging you belongs to a contact, the picture assigned to them plus their name is displayed. Contacts that you have designated as VIPs (see Chapter 5) have a little orange star displayed on their picture. If the number isn't in your contacts, the number is displayed.

The message list not only tells you a bit about who sent the message, but also previews the most recent message in a conversation (which might be the first message in the conversation). SMS messages have the first few words displayed, and MMS messages (text messages that contain a picture, video, or audio file) have a tiny preview of the attachment and the word *Attachment* in the preview area (**6.3**).

Scott McNulty 3:15 PM
Attachment

6.3 When you receive a picture, there's a small preview in the list.

Long tapping the picture to the left of a message brings up the quick contacts view (6.4). This allows you to quickly create a contact, add the number to an existing contact, or call the number with one tap and ask who the heck is texting you.

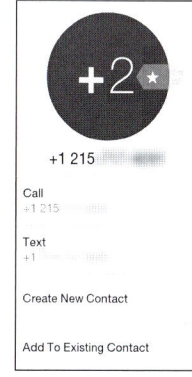

Long tapping the picture of a person already in your contacts brings up additional information about them and allows you to add a number or call their number. For either contact or contactless numbers, tap the star in this view to add this person to your VIPs.

Long tap a conversation to bring up the Delete menu (6.5). Tap Delete to delete all the messages in that conversation.

The Messaging app has no left panel, which is traditionally devoted to navigation, because the only navigation needed is the list of conversations. There is, however, a right panel. In the message list, swipe from the right edge to the left, or tilt the right edge toward you, to bring out the right panel (6.6). This panel displays all the pictures and videos you've received via MMS. Tap an image to see it full size. The picture displays, along with two icons at the bottom of the screen: download and share. Tap the download icon to save the picture to your Fire. Tap the share icon to bring up the Share panel.

6.4 Quick Contacts allows you to create a contact or add this number to an existing contact.

Swipe up from the bottom of the phone to return to the message list.

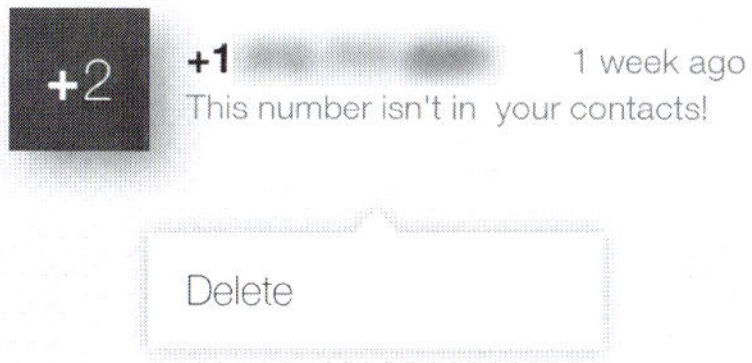

6.5 Long tap a conversation to bring up the Delete menu.

6.6 The right panel displays all the pictures you've received.

CHAPTER 6: MESSAGING   81

Sending a Text

There are a couple of ways to send a text message from your Fire:

- In the Contacts app, tap the contact that you want to text (or long tap their icon and bring up the quick contact menu). Then tap the label of the phone number you want to text (Home, Work, and so on).
- Long tap a picture in the Photos app, and tap Share. Select Messaging from the Share panel and a new message is created, with the picture in the message field.

You can also create a new message right from the Messaging app:

1. Launch the Messaging app.
2. Tap the Compose button in the upper-right corner.
3. The New Message screen appears (6.7). This screen has three major sections: the To field (where you address the message), the display section (which shows all the messages in this conversation), and the message section (where you type and send your message).

   To send a message directly to a phone number, tap the To field and enter the number. To type additional phone numbers, add a semicolon at the end of the number. The number you typed will turn orange, and you can then type another number.

   As you type a number in the To field, the Fire displays suggested contacts that have those numbers associated with them (6.8). Tap a contact to have your Fire fill in the rest of the number. You can also type a contact's name in the To field and your Fire will attempt to autocomplete it.

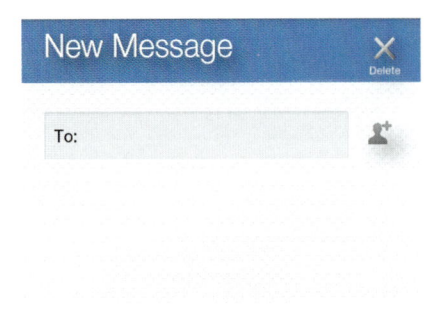

6.7 A blank message.

6.8 As you type, suggestions from your contacts appear.

If you'd rather choose from a list of your contacts, tap the icon to the right of the To field. VIPs are displayed first (once for every phone number they have in your contacts), with the rest of your contacts in alphabetical order. Tap a contact's name to add them to the message. Repeat the process to add multiple recipients.

Tap the name of your contact in the To field to see a list of other numbers you have for them (6.9). The number the message will be sent to has a checkmark next to it. If you want to send to one of the other numbers, tap it and the checkmark will appear next to that number.

4. Now that your message is addressed, you need to compose it. Tap in the field that says "Type message" and, well, type your message.

▶ **TIP** Tap the microphone key on the keyboard to dictate a text.

Tap Send and the message moves into the conversation field in a blue bubble with the word *sending* (6.10), which is then replaced with the time that the message was successfully sent.

If you want to send a picture or video, bring up the right panel while you're in a conversation. Tapping the camera icon allows you to take a new picture or video and inserts it into the text message. The rest of the panel displays all the pictures on your Fire. Swipe up and down to scroll through them, and tap one to select it.

You can also tap the camera icon next to the message field, which brings up the options Capture a Photo, Capture a Video, and Choose Existing (6.11).

6.9 When a contact has more than one number, you can select which one to message.

6.10 Sending a text, with love.

6.11 You can take a new photo or video to send, or choose an existing one.

Once the photo or video is selected or taken, it is inserted into the message field. Tap Send to send it.

Your text is sent!

Reading Texts

Reading text messages on your Fire is easy. Tap the Messaging app, and then tap the conversation you want to read. To reply to a conversation, type into its message field.

Long tap a message in a conversation, and you'll see a few options (6.12):

6.12 Long tap a message to reveal these options.

- Copy: This copies the long tapped message to the Clipboard so you can paste it into another text field or another text message.
- Forward: Tap Forward, and a new text message is created with the text of the message you selected, ready for you to send. All you have to do is add the recipients and tap Send.
- Delete: When you long tap a conversation in the message list and tap Delete, the whole conversation is deleted. When you long tap a message in a conversation and tap Delete, only the selected message is deleted from the conversation.

Long tapping an MMS message (a text message that includes a picture or video) reveals a menu with Save, Forward, and Delete options. Forward and Delete do the same things that they do in text messages, while tapping Save saves the picture or video to your Fire's Photo album. You can now delete the message or conversation, and still have the picture or video.

Find a message

You can search all of your text messages with a single tap:

1. In the messages list, tap the search icon.
2. A search field appears. Type a search term.
3. As you type, messages that include the term are listed below the search field, with a preview including the search term displayed in bold (6.13). Tap the message that you're interested in reading, and it opens.

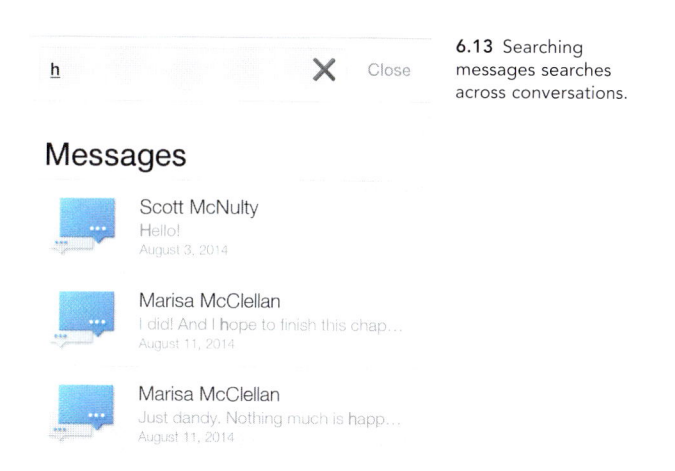

6.13 Searching messages searches across conversations.

Tap the X to clear the search term and look for something else.

Notifications

Sending text messages is fun, but it is even more fun to receive a message from a friend. Your Fire alerts you to new messages in several ways: a banner, an alert, and a vibration.

To customize these, follow these steps:

1. Bring up the Quick Settings panel, and tap Settings.
2. Tap My Accounts > Sounds & Notifications > Manage Notifications.
3. Scroll through the list of applications until you find Messaging, and tap it.

4. There are four options for text messages (6.14). Toggle off and on notifications, banners, and vibrate by tapping their on/off buttons. You can also set a custom sound by tapping the Sound section and selecting a new sound from the list. For more on these options, see the "Notifications" section in Chapter 4.

You can set a custom alert sound for any contact, much like you set a custom ringtone. Under Sounds & Notifications, tap Select Text Message Tones for Specific People. This brings up a contacts list indicating the text alert that is currently assigned to them (6.15). Tap a contact to select a different sound from the list.

6.14 The Messaging notification settings.

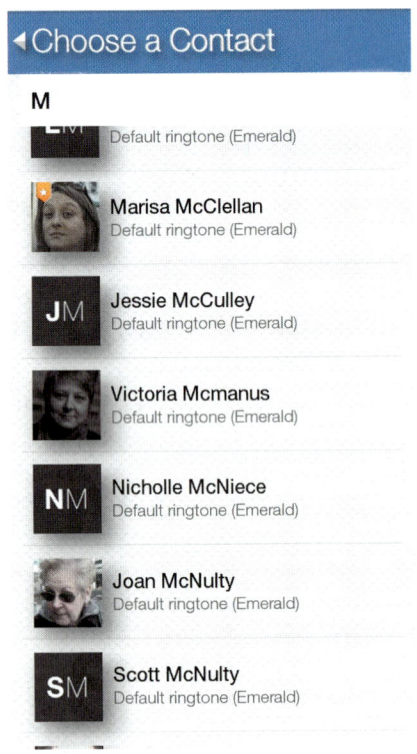

6.15 Your contacts and the text ringtone assigned to them.

86   THE AMAZON FIRE PHONE

CHAPTER 7

Phone

You already know how to use a phone, but the Fire has a few options that might not be obvious. This chapter covers making and receiving phone calls, conference calling, and responding to a call with a text message (which is the civilized thing to do, if you ask me; who talks on the phone anymore?).

Making a Call

Making a call with the Fire phone is simple. Tap the phone icon and you're greeted with a familiar keypad (7.1). Tap the keys to enter a phone number, and tap the blue Call button to initiate the call. If the number you're calling isn't in your contacts, a button will appear beneath the number (7.2). Tap it to create a new contact for that number.

Speaking of contacts, your Fire has all your contacts in it, so why manually enter a number? At the bottom of the phone screen you'll see change this to four icons. They are, from left to right, recent calls, contacts, keypad, and voicemail.

Tap the contacts icon to see a list of your contacts. You can scroll through them or search for a particular contact. Tap the contact, and you'll see all their numbers listed. Tap the number you want to call.

While you're in the phone app but not on a call, you can make quick calls to your VIP contacts by opening the right panel (swipe from the right edge to the left, or tilt the right edge toward you) and tapping the VIP you want to call (7.3).

7.1 The phone keypad.

7.2 When you call a number not listed in your contacts, an Add to Contacts button appears.

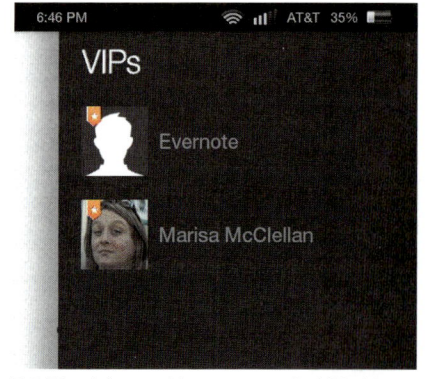

7.3 The right panel lets you to call a VIP with one tap.

While you're in a call, a photo of your contact is displayed (if there is one) along with a big red End Call button (7.4). Tap the red button to hang up.

At the top of the screen is the amount of time you've been on the call. Above the End Call button are four icons:

- Speaker phone 🔊: Tap this to use the speaker phone.
- Mute 🔇: Tap this to mute the outgoing audio.
- Keypad ⚌: This brings up the keypad in case you need to enter numbers.
- Add call ➕: Tap this to add another call to the one you're currently in (see the next section for details).

While you're in a call, the right panel has four quick links (7.5). Tap one to take a note, surf the web, check your calendar, or look at your contacts while you're on the phone. When you leave the phone app, a banner with green text is displayed at the top of the screen (7.6). This lets you know that you're still on the phone and how long you've been on the call. Tap it to return to the phone app.

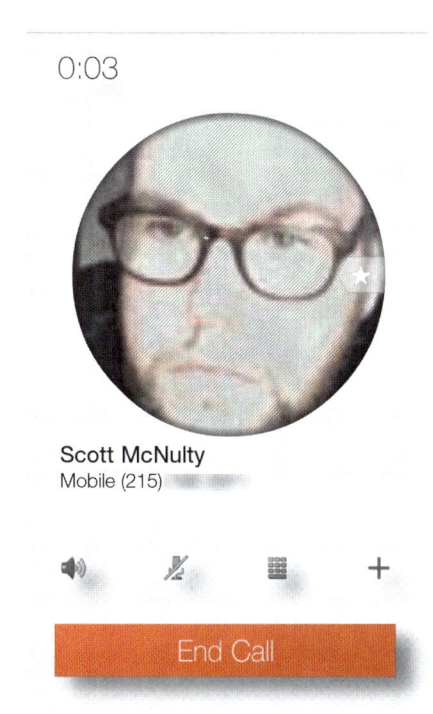

7.4 On a call. The call length is at the top of the screen.

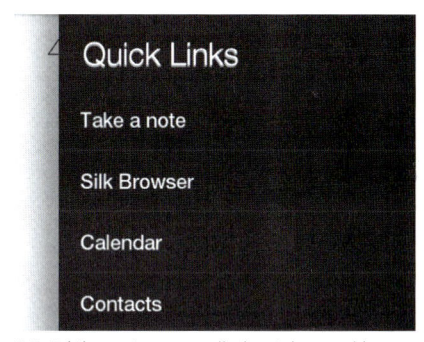

7.5 While you're on a call, the right panel has links to some handy apps you might want to consult.

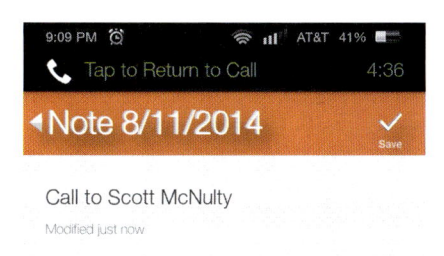

7.6 A banner with green text is displayed while you're on a call but in a different app.

CHAPTER 7: PHONE

Tap the Recent Calls button at the bottom of the phone app to see a list of all the calls you've made and received recently. Tap a call in the list to call that number or contact. Long tap to delete the call from the list.

> **TIP** Use Peek to see the precise date and time of the call and whether it was an incoming or outgoing call.

Conference call

One of the neatest features of the phone app is how simple it is to start a conference call:

1. You must be in a call to start a conference call. Tap the plus icon and select the contact you want to add to the call (see 7.4).

   That contact is called. Once the call is connected, a Merge button appears (7.7).

2. Tap the Merge button to conference both calls (7.8). Repeat with any other contacts you want to add.

Note that when you hang up, the conference call is over.

7.7 Tap Merge to create a conference call.

7.8 There are three people on this call: the Fire phone, Scott, and Marisa.

Answering a call

When your Fire rings, you answer it by swiping up on the screen (7.9).

If you don't want to talk on the phone, tap the call options button, which looks like a text message bubble, to reveal a right panel (7.10). If you tap one of the messages, the call will be declined and the caller will be texted the message you tapped. Tap the Answer button to accept the call, or tap Decline to decline it and send the call directly to voicemail. Pressing the home button on the Fire will silence the ringer and send the call to voicemail as well.

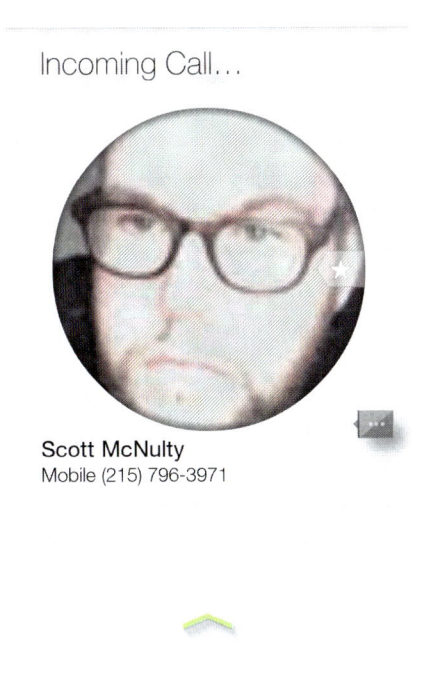

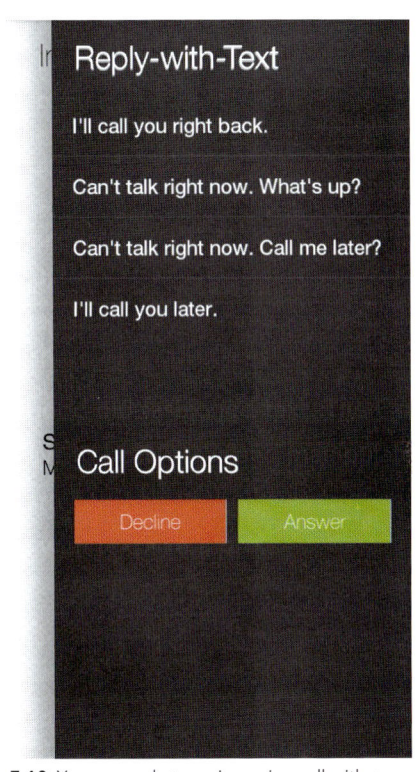

7.9 Slide up to answer an incoming call.

7.10 You can reply to an incoming call with a text, or decline it.

To add custom messages to the call options panel:

1. Bring up the Quick Settings panel, and tap Settings.
2. Tap Phone, and then tap Edit Reply-with-Text Messages.
3. The default messages are listed. Tap a message to edit it (7.11). You can change the messages completely, but keep in mind that you can't have more than four.

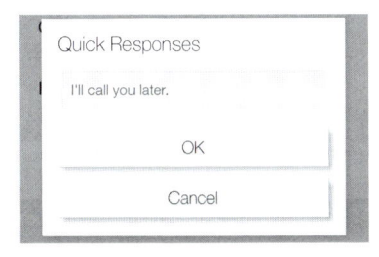

7.11 You can edit these, but you can only have a total of four.

CHAPTER 7: PHONE 91

You can also use the included headphones to answer an incoming call. The headphones have a microphone and a control fob built right in. When the headphones are plugged into the Fire's headphone jack, the audio will play through the earpods, so you'll hear whatever ringtone you've set. Press the button on the front of the control fob—it is the only raised button on the headphones—to answer the call. Press it again to hang up. You don't even have to take your phone out of your pocket!

Settings

You can set a custom ringtone on your Fire, but you can't use your own ringtones. For example, you can't use a song by your favorite band; only the included ringtones can be used.

To change your ringtone, and disable or enable vibration, follow these steps:

1. Bring up the Quick Settings panel, and tap Settings.
2. Tap Sounds & Notifications > Change Your Ringtone (7.12).
3. Tap the Sound section to select a new ringtone from the list of available ringtones (7.13). Tap a ringtone to set it and hear what it sounds like.
4. When you're happy with the ringtone, tap the checkmark in the upper-right corner.
5. Tap On/Off in the Vibrate section to enable or disable your phone vibrating when you get a phone call.

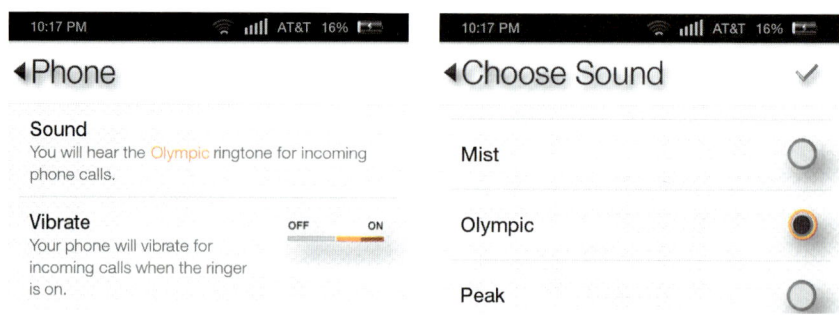

7.12 The phone notification options.

7.13 The list of ringtones available on your Fire.

Volume

To raise or lower the volume, use the volume buttons on the side of your Fire (see Chapter 1 for more information). When you press one of the volume buttons the Ringer Volume slider appears (7.14). You can use the buttons to move the slider up and down, or you can slide your finger along the bar to fine-tune the volume.

Underneath the Ringer Volume slider are the ringtone controls. By default, it is set to Ringer On. Slide it to Vibrate to silence the ringer but enable vibrations. Silent turns off the ringer and the vibration, and Silent for 3 Hours does the same thing but only for three hours (plus it displays a countdown right in the controls) (7.15).

7.14 Ringer volume can be controlled with the slider or the volume buttons.

7.15 Setting your ringer to silent for three hours starts a countdown.

You can use the headphones to control the ringer. The control fob on the headphones has a + button (on top) and a – button (on bottom). Click the + button to increase the ringer volume. Click – to lower the volume, until the ringer is off. Once that happens, the Fire switches to vibrate only. Press the – button once more to put it in silent mode. Press + once to return it to vibrating and ringing.

▶ **TIP** While you're on a call the same methods control the call volume.

Additional settings

There are a few phone settings that you might be interested in. To get to them, go to Quick Settings > Settings > Phone. Once you're in the Phone section, tap any entry; they all take you to the same place (7.16).

- Call Waiting: On by default. If you are on a call and get another, you can put the first on hold to answer the other. Tap Off to turn this off (which will give the second caller a busy signal).

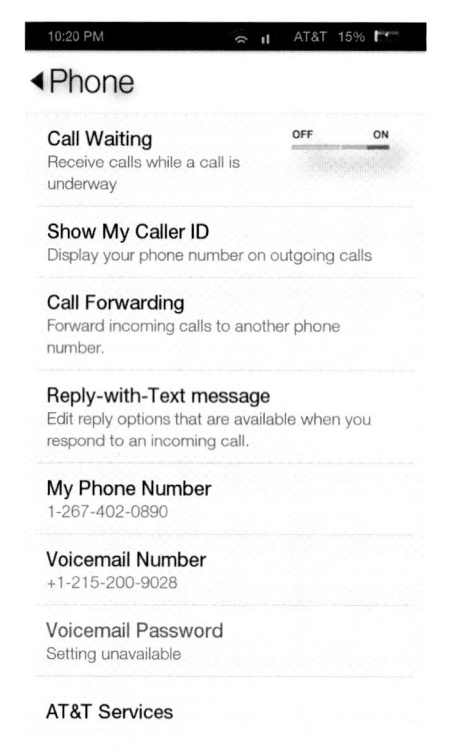

7.16 Phone settings.

7.17 You can forward calls from your Fire to another number.

- Show My Caller ID: This setting controls how your phone number appears on other people's phones. Tap to see the three options: Network Default, Hide Number (if you don't want people up in your business), and Show Number.

- Call Forwarding: You can forward calls from your Fire's number to another number. Tap this setting, and then tap On to provide a number to forward to (7.17). Tap Off to disable call forwarding.

- Reply-with-Text Message: Covered in the previous section.

- My Phone Number and Voicemail Number: These are for your information.

- Voicemail Password: If visual voicemail is enabled, you can change your password by tapping this.

- AT&T Services: Tap this to get a list of numbers you can text and call for information about your AT&T service.

CHAPTER 8

Calendar

If it isn't on my calendar, it doesn't exist, which is why I make heavy use of the calendar on the Fire. It can sync to your Google, Exchange, and Yahoo calendars, and others. You can also just keep your calendar events on your Fire if you aren't a syncing kind of person. This chapter covers viewing calendar events, creating and inviting people to events, deleting unwanted events, navigating between multiple calendars, and more.

Setting Up Your Calendar

When you add an account to your Fire that supports calendar syncing, it is automatically synced. You can turn off calendar syncing by going to Quick Settings > Settings > My Accounts > Manage Email Accounts. Tap the account, and then toggle Sync Calendar to off. The events associated with that calendar will be removed from your phone, though they aren't deleted from the calendar itself.

Facebook

If you use Facebook, you can sync your Facebook events to your Fire phone. Under Manage Email Accounts, tap Calendar Settings (8.1). The Sync Facebook Events setting is off by default, so tap it to toggle it on. If you haven't linked your Facebook account to your phone, you'll be prompted to do so.

The Fire asks for permission to access your events on Facebook (8.2). Tap Allow, and after a few seconds your Facebook events will start showing up on your phone.

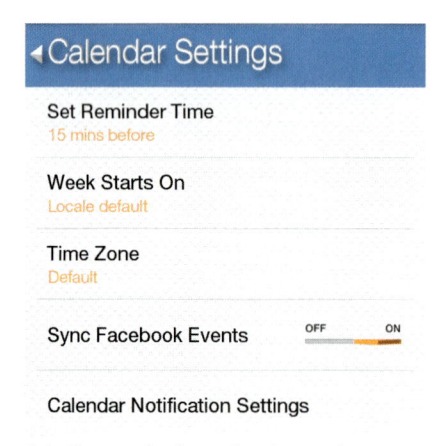

8.1 All your calendar settings in one place.

8.2 If you're a big Facebook user, you'll want to sync Facebook events to your phone.

Viewing Your Calendar

Launch the calendar by tapping its icon in the home grid or on the Carousel. On the Carousel, the Calendar app displays the current date (8.3). Under the icon, you'll see a list of upcoming events color-coded by calendar. Only future events are listed here, and you can swipe up and down to scroll through them. Each entry lists the time, title, location, and length of the event. Tap an event and you'll be taken to its details.

The calendar opens to the today view (the default view) at the current time (8.4). Each event shows a title and location. Since I have more than one calendar synced to my Fire, the events are color-coded: orange events are from my Google calendar and blue from my Exchange calendar.

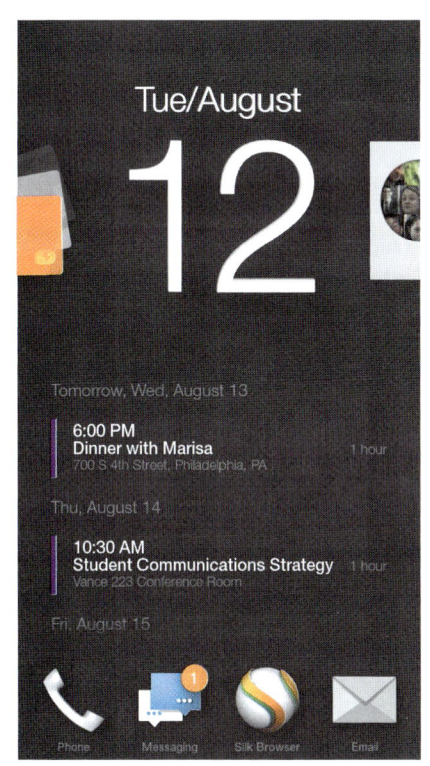

8.3 The calendar icon in the Carousel displays upcoming events.

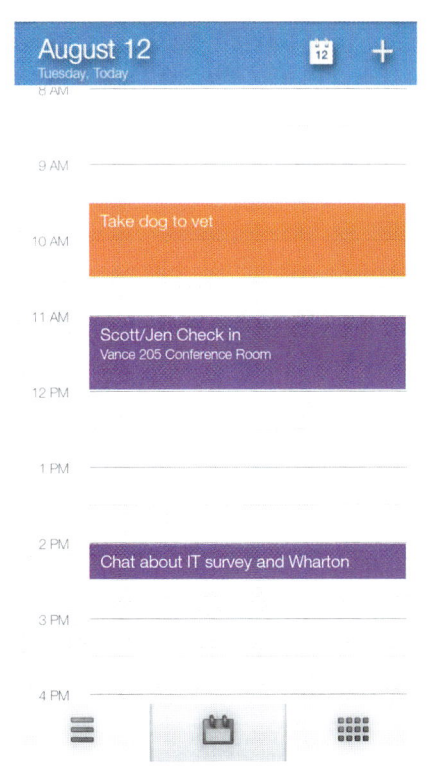

8.4 Each event is color-coded according to the calendar it is on.

CHAPTER 8: CALENDAR

8.5 The left panel lists all the calendars that are syncing to your device. Uncheck a calendar to hide the events on it.

8.6 The right panel lists quick messages you can send to the attendees and organizers of your next event.

Open the left panel to see all the calendars you're currently syncing and the colors assigned to them (8.5). You can display or hide events on a per calendar basis. Tap the box next to the calendar you want to hide, and the checkmark is removed and the events are no longer displayed. The events aren't deleted from your calendar, nor is the calendar removed from syncing to your Fire; the events just don't show up. To display them, tap the box in the left panel once more and the events reappear.

The right panel displays the next event scheduled on your calendar, along with some quick messages you can send to people attending it (8.6). (I talk more about these quick messages later in this chapter.)

Along the bottom of the calendar screen, you'll find three icons. Each gives you a different view: list, day, and month.

- List view: Tap the list view icon to see all your events in a list (8.7). Each event has a color bar next to it, indicating which calendar it is from. Swipe up and down to scroll through your past and future events.

    Each event shows you the time it starts, the name of the event, the location, and, at the far right, the event length. Using Peek (tilt the phone), the location is replaced with the organizer of the event. Tap an event to see its details.

- Day View: Tap the day view icon to see all your events for a particular day; this is the default view when you launch the calendar. Swipe right to see days in the past, and swipe left to see future days.

- Month View: Tapping the month view icon gets you a monthly overview of your calendar (8.8). Swipe left or right to change the month displayed. If a day has any events scheduled, colored lines representing the calendars the events are on displays. No matter how many events are scheduled for the day, only one line per calendar displays.

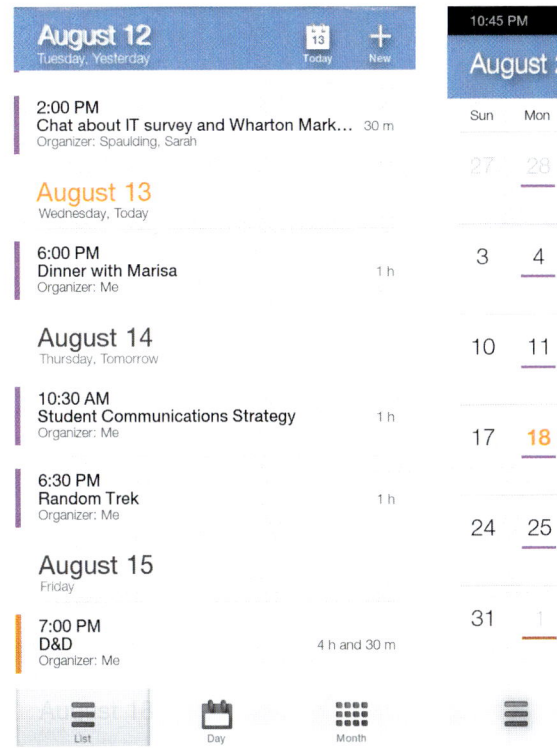

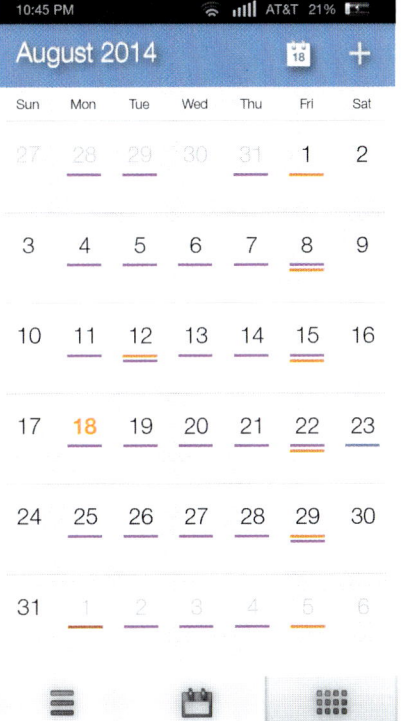

8.7 This list of events is a great way to see all the events that are coming up on your calendar.

8.8 The monthly view. A colored line on a day means there is at least one event on that day.

CHAPTER 8: CALENDAR   99

Tap a day to switch to the day view and see the events on your calendar for that day.

▶ **TIP** In any view, you can quickly return to the current day by tapping the today icon at the top of the display.

There is an additional calendar view that doesn't have a button assigned to it. If you hold your Fire in landscape mode (that is, turn it horizontal) in the calendar, you'll see a five-day view (8.9). Swipe up or down to go through the hours of the day and see what events are happening on the timeline. All-day events are displayed at the top of the screen under the day. To see another five-day period, swipe left or right. To return to one of the other views, turn the Fire back to portrait mode.

8.9 Turn your phone to landscape to see the five-day view.

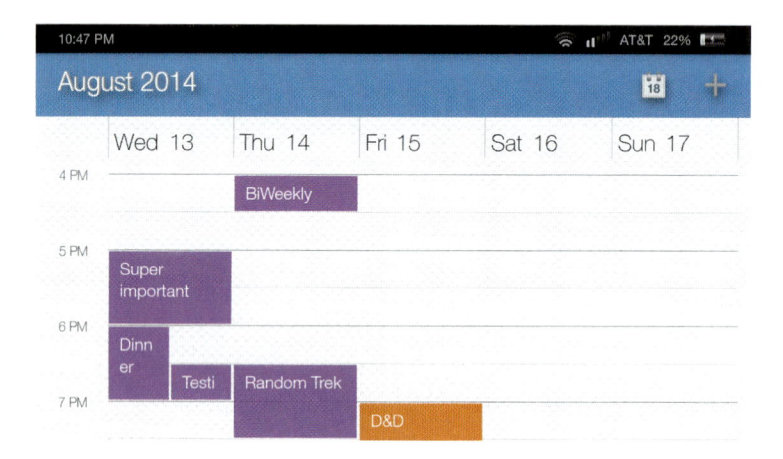

Events

When you tap an event on a calendar you're taken to the event details (8.10). The title of the event is listed at the top, and under that you'll find its date, time, and location. Tap the location, and Maps will launch and try to find the location for you (if the location is a conference room, as it is in 8.10, Maps won't be able to locate it unless you're meeting in a very famous conference room). To round out the details, the organizer of the meeting is listed. You're the organizer of any meetings you created. If someone else is the organizer of an event, you can tap their name to email them.

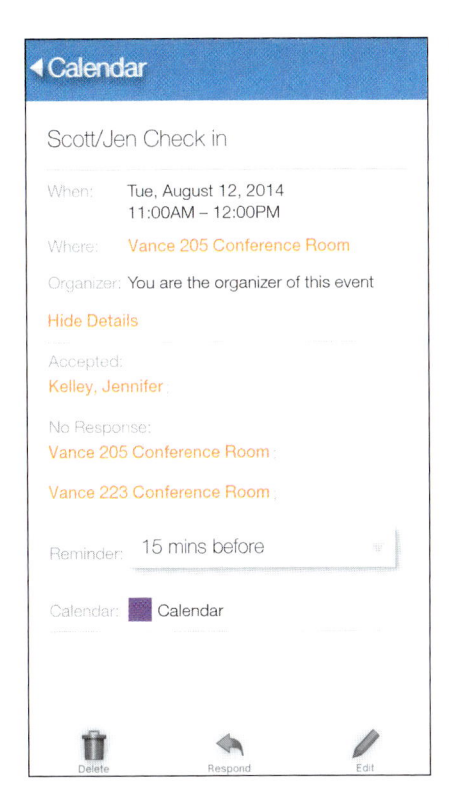

8.10 Event details include attendees.

The More section (which can be displayed by tapping More) displays additional information about the event. If this is an event that you've organized and invited other people to, you can track responses here. The meeting reminder and the source calendar round out the More section.

There are three action buttons at the bottom of every event for which you are the organizer:

- Delete: Tap this button to delete the event. An alert pops up making sure you want to delete the event. Tap OK to delete; tap Cancel to keep it.

- Email: Tap the Email button to reveal three options: Reply, Reply All, and Forward. Tapping any of those options creates a new email, but each one does so differently. Reply creates an email addressed only to the organizer of the event and uses the title of the event as the subject of the email. Reply All addresses an email to the organizer and anyone who is invited to the event (and, once again, uses the title of the meeting as the subject of the email). Forward creates an email with the event details in the body of the message, and the title of the event as the subject, but it isn't addressed to anyone. Type the email address you want to forward this information to.

- Edit: Tap the Edit button to change the details of an event for which you are the organizer (8.11). These fields are covered in detail in section "Creating an event." Tap the checkmark in the upper-right corner to update the event, or tap the X to discard your edits. Keep in mind that if there are additional guests (other than yourself) associated with the event, they will receive an update to the event. There is no confirmation, so be sure you want to update the event.

If you are invited to an event, you'll get a notification in your email (8.12). Tap either the checkmark (accept) or the question mark (tentative) to add the event to your calendar. Tentative events are marked with a color bar with slashes to indicate you aren't sure if you're attending. Tap the X to decline the invitation.

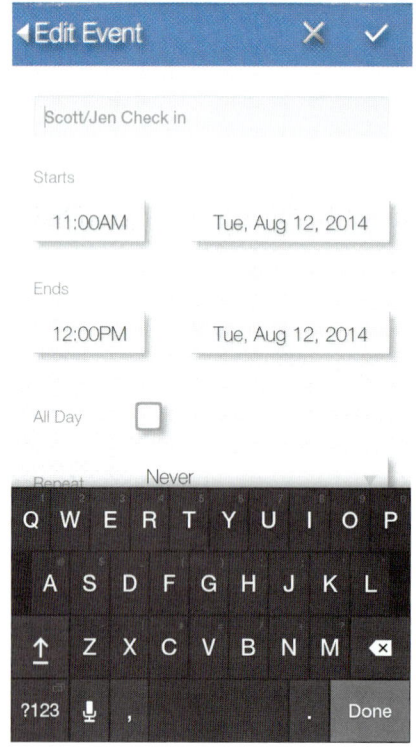

8.11 Editing an event.

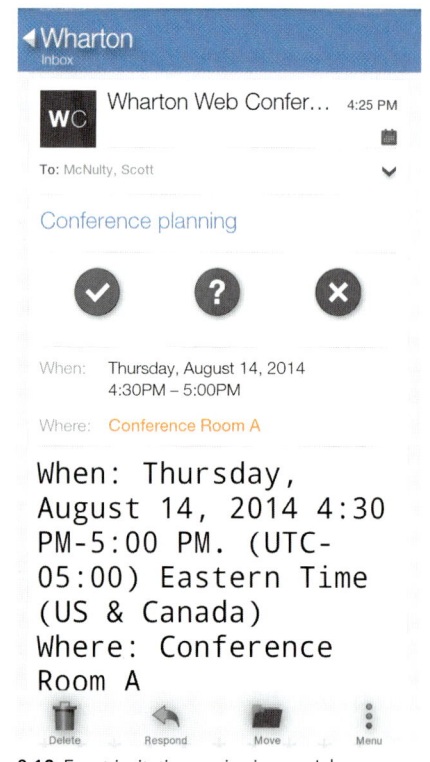

8.12 Event invitations arrive in your Inbox.

102   THE AMAZON FIRE PHONE

Once the event is added to the calendar, tap it to see its details. But you won't see an Edit button at the bottom of the screen. If you aren't the organizer of an event, you won't be able to edit it.

The right panel

When you're in the details of an event, the right panel has some additional actions you can take, depending on the kind of event.

- An event with no guests: This is an event that you created but didn't invite anyone to. Bring up the right panel, and your VIPs and recent contacts are listed (8.13). Tap any or all of them to add them to the event. Your Fire sends out an email invitation to each of them.

- An event with guests: The right panel lists eight quick messages that you can send to all guests (8.14). Tap one to create a new email with the subject "Re: [the event title]" addressed to all the guests. The quick message you tapped is in the body of the email. All you have to do is tap Send.

- An event in the past: This right panel is the least exciting. It tells you that the event ended x minutes/hours/days ago and lets you email the organizer if you were a guest.

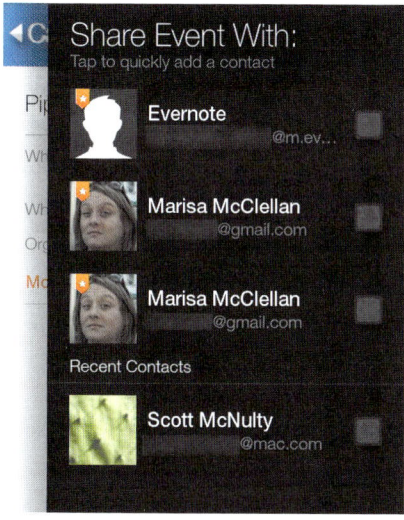

8.13 Add invitees to events using the right panel.

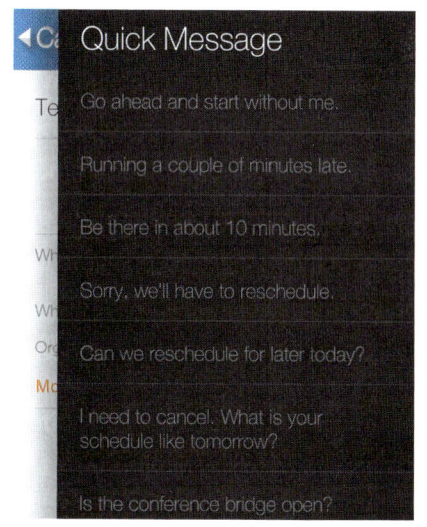

8.14 Quick messages let you send canned emails to the organizers or guests of events.

Creating an event

To create an event on your Fire, open the calendar app and tap the + button in the upper-right corner. The New Event screen appears, awaiting your input (8.15).

Every event has the same basic makeup, and you can fill in as little or as much of it as you like:

- Title: This is displayed on your calendar and should be related to whatever the event is. For example, you might create an event titled Dinner with Mom.

- Starts: Set the time and date for the start of your meeting. Tapping the time brings up the time selector (8.16). Swipe up or down to set hour, minute, and AM/PM, and then tap OK. Tap the date to bring up the Set Date screen.

- Ends: Once you set the start date and time, the end date and time will be set to one hour later. To change it, to just tap the time/date and enter the ending time/date you'd like.

- All Day: If your event is going to last all day, tap this box. The Starts and Ends fields change to dates. Keep in mind that you can create an all-day event that spans multiple days by selecting different starting and ending days.

- Repeat: There are a variety of options to have your Fire automatically repeat events on your calendar. Tap the Repeat menu, which defaults to Never, to see the full list (8.17).

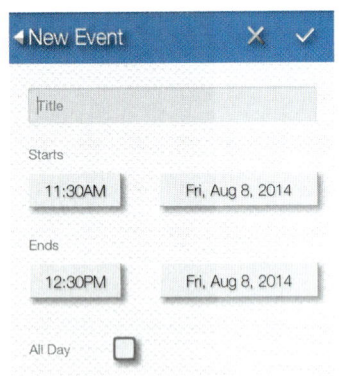

8.15 Creating a new event.

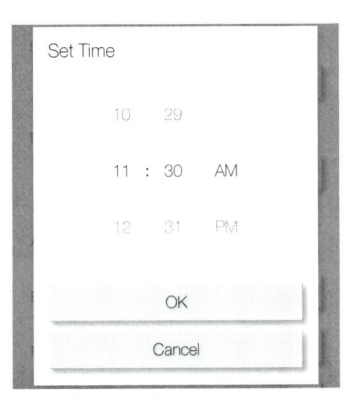

8.16 The time selector.

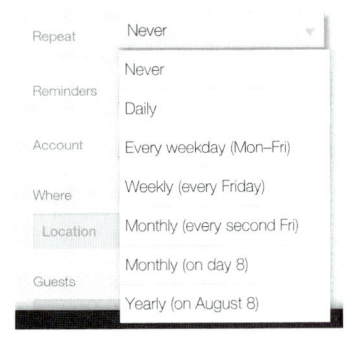

8.17 You can create events that repeat with a variety of schedules.

- Reminders: By default, every event you create will remind you (and anyone who accepts an invitation to that event) 15 minutes ahead of time. Tap the Reminders menu to see the full list of options (including None and At Start of Event) (8.18).

- Account: Select the calendar on which you want this event to be created. If you're inviting people (more on that in a second), this will also determine which email address the invitation is from. Only calendars that you have synced to your phone will appear in this list, color-coded in the same manner as events on the calendar.

- Where: Enter a location for your event. You can enter whatever you like here: an address or the name of a conference room, for example. Just be sure that the people you invite people will know what you mean by whatever you enter (assuming you actually want them to show up).

- Guests: Entering guests into an event will email invitations to those people. You can simply start typing, and suggestions will appear based on who is in your contacts (8.19). If you're inviting someone who isn't in your contacts, enter their full email address. As soon as you finish typing the email address and hit the spacebar, the address will turn orange and a semicolon will appear. This means you can start adding more people to the guest list. You can also add people directly from your contacts by tapping the Add Contact button. This brings up your contacts list, where you can pick a contact to add. Repeat this process until you've added all the people you want to invite.

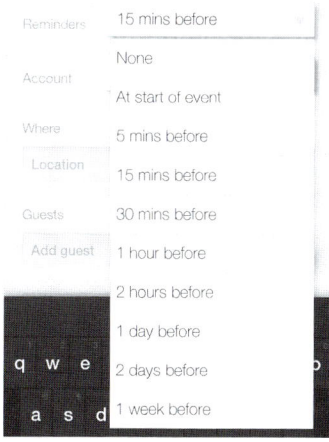

8.18 Reminders aren't just for you; they also appear on guest's calendars.

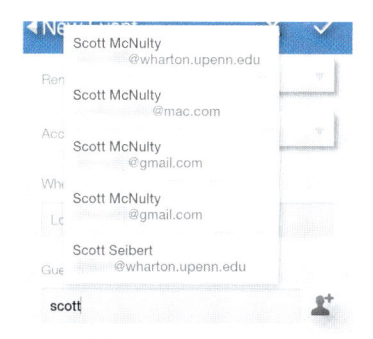

8.19 Adding a guest to an event searches your contacts.

CHAPTER 8: CALENDAR   105

- Notes: This is a great place to include additional information about the event that you want to remember (like hotel confirmation numbers, perhaps) or that you want attendees to know (the agenda for the meeting).

Once you've finished adding details, tap the checkmark at the top of the screen to save the event to your calendar and send out the invitations. Tap the X to cancel the process.

You can also create an event by tapping an empty hour slot on your calendar. A New Event icon appears in that time block (8.20). Tap it, and you're taken to the same event creation screen, with a key difference: the date and time are filled in for you based on where you tapped the screen.

8.20 Creating an event is as easy as tapping a time.

Responding to an event invitation

When you are invited to an event, it will appear on your calendar with a striped banner on it. You'll also receive an email.

To respond via the calendar:

1. Open the calendar.
2. Tap the event.
3. Three icons appear. From left to right, they are Accept, Tentative, and Decline (8.21). Tap one and you're done. The Fire will send the proper response to the event organizer and either add the event to your calendar or remove it.

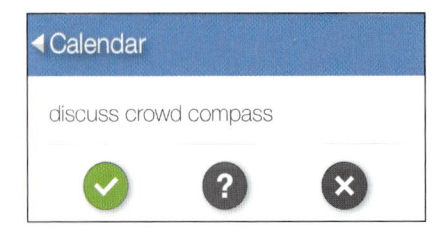

8.21 When you're invited to an event, you can accept it, tentatively accept, or decline.

Settings and notifications

You can determine how the calendar alerts you to invitations and upcoming events.

To get to the calendar settings:

1. Open the calendar.
2. Bring out the left panel, and tap Settings in the More section.
3. Tap Calendar Settings.

The first thing you can set is the default reminder time. Tap Set Reminder Time, and you'll see the same options as you have when setting a reminder time in an event: anything from None to 1 week before the event.

Week Starts On determines when the week starts in the calendar display. By default, it is set to Locale Default. Tap it, and you can set it to Saturday, Sunday, or Monday (8.22).

Time Zone determines the time zone that your calendar will display events in, and more importantly, the time zone used for sending events to others. Tap the time zone to change it from the default (which is set based on your location).

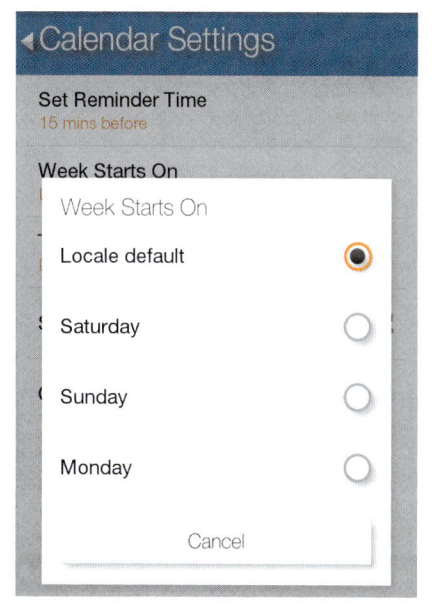

8.22 When do you want your week to start?

The calendar notification options are the same as the other notifications on your phone: Notifications, Banners, Sound, and Vibrate. By default, they are all enabled. Tap any one to disable it. Tap Sound to change the alert sound or set it to None.

When an event reminder goes off, your phone will vibrate, make a sound, and show a badge on the lock screen (8.23). The number next to the calendar icon indicates how many event reminders have gone off. Unlock the phone and you'll see alerts in the notifications area of Quick Settings (8.24).

You can dismiss the notification by swiping left or right across it. However, that means you won't be alerted about the event again. If you just want to snooze the alert for five minutes, swipe down on the three lines in the center of the notification and tap Snooze. You'll be alerted again in five minutes so you won't be late for your meeting.

8.23 A calendar notification on the lock screen.

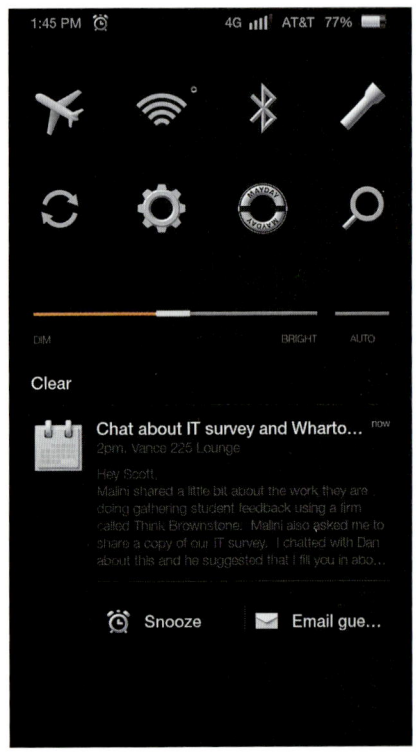

8.24 An expanded calendar notification on the Quick Settings screen.

108   THE AMAZON FIRE PHONE

CHAPTER 9

Silk

The popularity of smartphones comes down to two things: apps and the ability to surf the web wherever you might find yourself (I have a friend who has dubbed this "having the Internet in your pants"). Your Fire phone has a built-in web browser called Silk, which Amazon designed from the ground up (with a little help from some open source code). You might think that Silk is just like any other browser, but this chapter tells you why that isn't the case and how to use its features.

What Makes Silk So Different

Amazon looked at existing browser technology and thought, "we can do better." They wanted to make browsing as fast as they could, but how could they do that without being able to control the bandwidth speed of the phone? The solution they came up with is pretty clever.

You might know that Amazon isn't just in the book, phone, and tablet business. They are also a big player in cloud computing. That's just a fancy way of saying Amazon has lots of computers that they manage and which they allow other business (like many of the most popular websites in the world) to use for a variety of purposes. Amazon gets to charge for this service, and the businesses don't have to manage their own computers. Everyone wins.

What does this have to do with Silk? Well, Amazon's engineers took a look at all that cloud computing power at their fingertips and decided to bake it into Silk with something called *cloud features*.

Cloud features preprocess websites to speed up browsing, and they even guess at the links you might visit and predownload them. But this means that your web traffic is going through Amazon's servers, which some people don't like. See the "Settings" section of this chapter for more about how these features work, why you might want to turn them off, and how to do so.

Visiting a Webpage

Tap the Silk icon to launch Silk (9.1). By default, Silk is on the row of icons on your home screen, but you can also find it in the home grid of apps if you've moved it.

9.1 The Silk icon.

When you launch Silk, it will show you one of two things:

- A list of your most visited pages (9.2). Tap any site in the list to visit it. If this is the first time you're using Silk, or you've cleared your browsing history (more on how to do that later in this chapter), you'll get a message saying that your history is empty.

- If you've switched out of Silk while browsing a website, that site will reload when you go back into Silk.

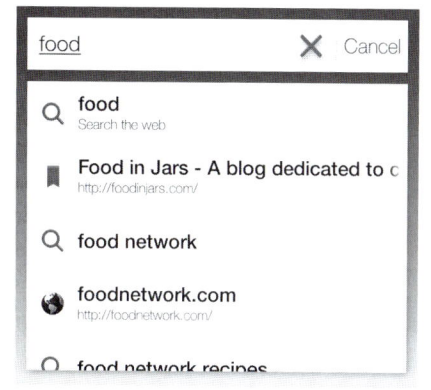

9.2 Silk shows you the sites you've visited most.

9.3 Type in a URL or a search term and Silk knows what to do.

▶ **TIP** When Silk is in the Carousel, it displays your recently visited websites. Tap one to go right to it.

At the very top of the screen you'll see a box with a magnifying glass icon. You can type either a URL or a search term here. As you type, Silk suggests things (9.3). Next to each result, an icon tells you what type of result it is:

🔍 : Tap a suggestion that displays this icon, and you'll perform a web search for that phrase using whatever search engine you've set (the default is Bing.com, but this can be changed; see the "Settings" section in this chapter).

🕒 : Your browsing history is also searched as you type, and if you've typed a similar word or phrase before, this icon shows up next to the result. Tap it to initiate a web search.

🌐 : Silk searches for web addresses that match what you're typing. Tap one to go right to the website.

🔖 : As with most other browsers, you can bookmark, or favorite, sites that you often visit. Tap to go directly to the website.

If none of the suggestions are what you're looking for, type your search or web address and tap the Go key. If you typed a search term, you'll be taken to the search results (9.4). If you typed a web address and the site is up, it will load. The URL bar will fill with yellow as it is loading the site. Tap the X to stop loading the site. Tap the refresh button to load the site again.

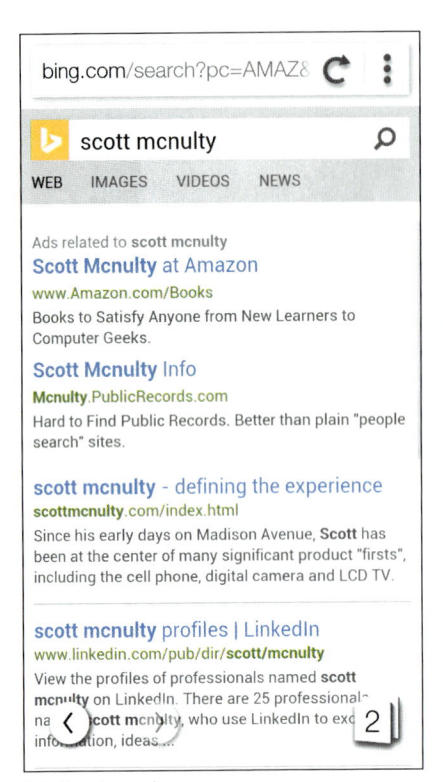

9.4 Search results.

9.5 Notice the arrow at the bottom of the screen. That's the auto-scroll arrow.

To scroll up or down on a page, you have two choices:

- Swipe up or down, as you would normally do.
- Tilt the top of your phone back until you see the auto-scroll arrow appear toward the bottom of the screen (**9.5**). The page will start scrolling down. The more you tilt the top of your Fire back, the faster the scrolling will happen (and the icon of the arrow will lengthen to show that you're going faster). To stop scrolling, tilt the top of the phone forward; to scroll up, keep tilting the phone forward until you see the auto-scroll arrow icon appear toward the top of the screen (and pointing up). Tilt forward more to scroll faster. As you reach the bottom or top of a page with auto-scroll, the arrow icon flattens out to a line, indicating that you can't scroll anymore.

▶ **TIP** Auto-scrolling is cool, but some people find it annoying. Turning it off is covered in the "Settings" section of this chapter.

9.6 Philly.com's desktop site.

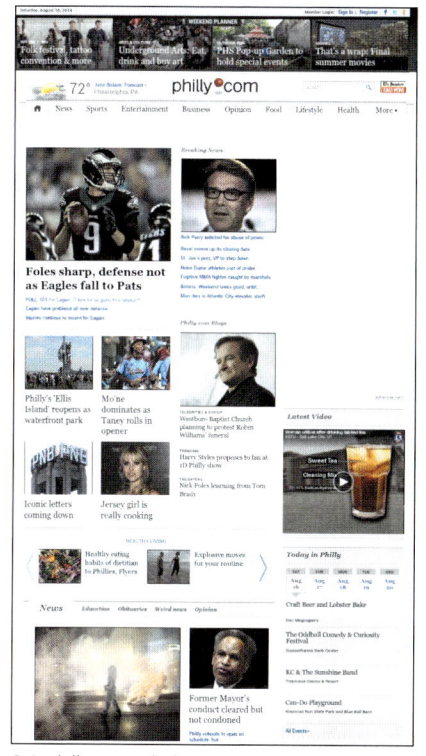

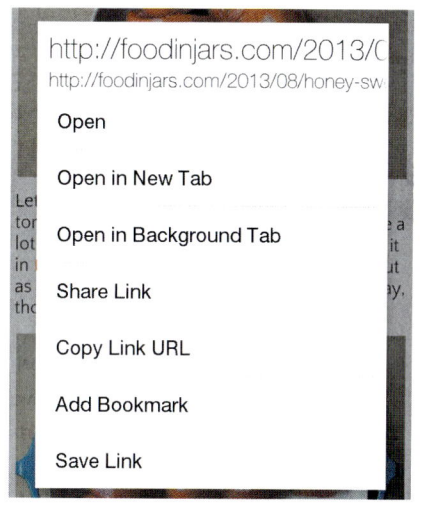

9.7 Long tap to see the link options.

Silk is a full-fledged browser, which means it is capable of displaying the full (or desktop) versions of websites. But if you're going to larger, well-trafficked sites, most of the time you'll get a mobile-friendly version of the site that is easier to read on smaller screens.

If a site doesn't offer a mobile version and you're presented with multiple columns of content, you'll quickly notice that is hard to read that text on your Fire's display **(9.6)**. Pinch to zoom in to a particular section of the site. You can also have the Fire auto-zoom onto a column of text or a photo by double tapping. The Fire will zoom right onto the doubled tapped text or photo, making it take up the whole display.

Links work just as they do in any browser: tap a link to follow it. Long tap a link to bring up a menu that displays the link URL at the top and a few options for you to tap **(9.7)**:

- **Open**: Tap Open to follow the link in the current browser. This is just like tapping the link on a page, but using this method you get a preview of where the link points to (so you can be sure you're not being directed to a site you don't want to visit!).

- **Open in New Tab:** This option opens a new tab, loads the page, and switches to that tab.
- **Open in Background Tab:** If you don't want to read the link you're opening immediately, this is the option for you. Silk will continue to display the page you're currently on, while opening the link in the background in another tab.
- **Share Link:** Tap this and the Share panel appears, listing the apps on your phone to which you can send the URL. This allows you to tweet, email, or text URLs without having to copy and paste them.
- **Copy Link URL:** If you would rather copy and paste, tap this, which copies the URL onto the Clipboard for pasting elsewhere.
- **Add Bookmark:** Bookmark a link for visiting later. (See the "Bookmarks" section of this chapter.)
- **Save Link:** Tap this to save the linked page or file to your Fire. Opening a saved page does not require a network connection, because it is saved on your phone.

Long tapping an image on a site brings up a menu with two options: Save Image and View Image. Save Image saves the image to the Photos app, and View Image opens the image in a new tab. It is possible for an image to be a link to another page. Long tapping an image that is also a link opens a menu with all the link options and the two image-specific options.

As you scroll through a site, you'll notice that the URL/search box is hidden to give more display space to the content you're looking at. Swipe your finger down no matter where you are on a site to bring the top interface back into view. When it is visible, you can see the URL of the page you're on. Tap in the box, and the address is highlighted and an X appears (9.8). Tap the X to clear the URL so you can enter another term. If you want to edit the URL you're currently on, tap the URL to deselect it and then place your finger where you'd like to start typing (or deleting). Tap the Cancel button to return to the site you were viewing.

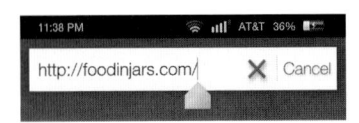

9.8 Tap the URL bar to edit it.

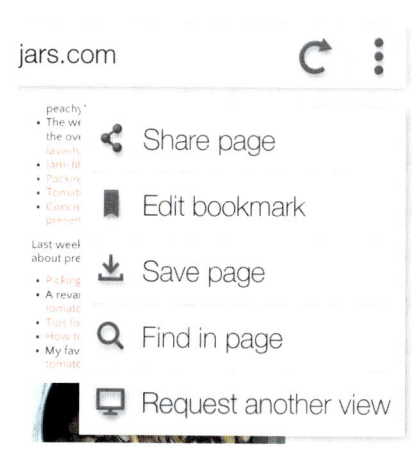

9.9 The menu button has a few options for you.

9.10 Search for terms within the page. Matches are highlighted.

To the right next of the refresh button is the menu button. Tap it to bring up a number of things you can do on any site (9.9).

- Share Page: Tap this to email, tweet, or share the page using an app installed on your Fire.

- Add Bookmark and Edit Bookmark: Tap to either add a bookmark for the page or to edit its bookmark.

- Save Page: Tap to download the site to your Fire. An alert is shown at the bottom of the display when the save process begins and again when the page has been saved. Keep in mind that it creates a *copy* of the site on your phone, so if the site gets updated, the copy on your phone won't reflect the update.

- Find in Page: Looking for a particular word or phrase in the site? Tap the menu button and then Find in Page. A search box appears in which you can type the word or phrase you're looking for. As you type, anything that matches is highlighted. Once you're finished typing your search string, all of the instances are highlighted in yellow, with the currently selected one highlighted in orange (9.10). The total number of matches is displayed next to the numbered instance that is currently selected. Use the up and down arrows to hop to other instances of the word or phrase in the page. While search is active, you can still interact

CHAPTER 9: SILK   115

with the page as normal: scroll, tap links, and so on. If you want to close search but stay on the same page, tap the X. When you follow a link away from the page being searched, the search bar is automatically dismissed.

- Request Another View: Many websites detect the size of your screen and then serve up a design that makes sense for the display (this is an oversimplification, I know, for all you web people out there). This means your Fire will generally get a site's mobile version, which shows less than the full, or desktop, version. By default, your Fire displays whatever version the site thinks is best. Tap Request Another View to request either the mobile or desktop version (9.11).

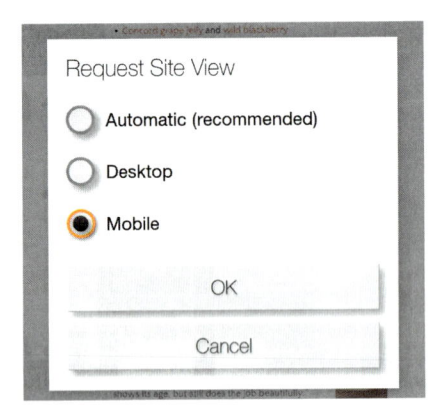

9.11 You can request different views of sites.

Keep in mind that some sites, especially personal sites or those from smaller businesses, have one look whether you're visiting on a phone, your desktop computer, or an Internet-enabled fridge. You can still use this function to request a different view, but you'll just get the same one because that's all there is to get!

At the bottom of the screen, you'll see a few icons (9.12). The back and forward arrows work exactly as they do in any other web browser. If you follow a link from one page to another, tap the back (left-pointing) icon to get back to the first page. Once you're back on the original page and you want to return to the other page, tap the forward (right-pointing) button.

The other icon tells you how many tabs you have open. Check out the "Tabs" section for more about this feature.

9.12 The back/forward and tab buttons.

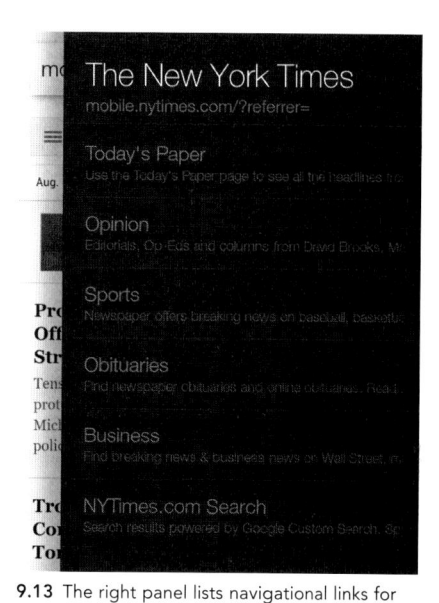

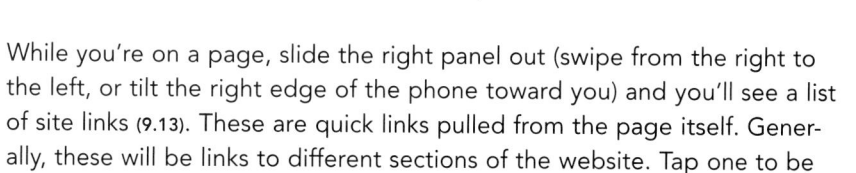

9.13 The right panel lists navigational links for many sites.

While you're on a page, slide the right panel out (swipe from the right to the left, or tilt the right edge of the phone toward you) and you'll see a list of site links (9.13). These are quick links pulled from the page itself. Generally, these will be links to different sections of the website. Tap one to be taken to that page. If a page doesn't have any site links, the right panel won't display anything.

Reader mode

Silk's Reader mode makes it easier to read articles on your phone. When you're looking at an article in Silk, a new icon appears in the top bar: a pair of glasses (9.14). This is the Reader mode icon. Tap it and Silk displays the article text in full screen (9.15). Now you can just concentrate on reading, without being distracted by links, ads, or any of the other stuff publishers put on their article pages.

9.14 Tap the glasses to get into Reader mode.

CHAPTER 9: SILK   117

9.15 Ads and other clutter, begone!

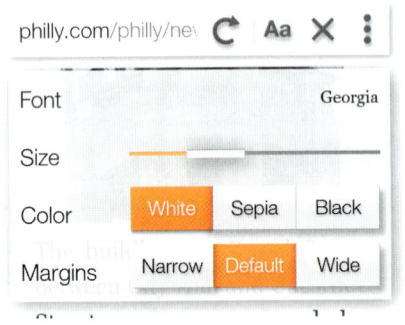

9.16 Reader Mode options.

You can customize how text is displayed in Reader mode. Tap the **Aa** button in the top bar to bring up the options (9.16). Change the font by tapping Font and choosing Georgia, Palatino, Baskerville, Helvetica, or Lucinda.

Use the Size slider to increase or decrease the font size. The Color buttons change the color of the background and text (White is black text on a white background, Sepia is dark sepia text on a light sepia background, Black is white text on a black background). The Margins buttons let you set how narrow you want the column of text to be.

The changes are applied immediately, so feel free to fiddle around with different combinations to find the one that works best for you.

To exit Reader mode, tap the X in the top bar (see 9.15).

Tabs

You might be familiar with the concept to tabs from the browser you use on your computer. Tabs are a way of having multiple websites open at once, but you only look at one tab at a time.

For example, say you're looking at a movie review site. As you scroll down, you note a bunch of reviews you'd like to read. You could tap each link and read them individually. Or you could long tap a link to open it in a tab in the background, move on to the next one in the index, and repeat the process until you have all the reviews you want to read in separate tabs.

The tab icon (9.17) shows how many tabs you have open at the moment. If you only have one tab open, that is the page you're looking at.

To see all the tabs you have open, or to create a new tab, tap the tab icon. This takes you to the tab grid (9.18). At the top of the tab grid you'll see the number of tabs you have open and an arrow. Tap the arrow to return to the tab you were just looking at.

9.17 The tab button tells you how many tabs you have open.

9.18 The tab navigator.

CHAPTER 9: SILK   119

Swipe up or down to scroll through the list of tabs. Tap any thumbnail to switch to that tab, and then tap the tab icon to come back to the tab grid.

To close a tab, tap the X in the upper-right corner of a tab preview. The tab closes without a warning, so make sure you really want to close a tab before you tap.

To open a new tab, tap the + sign in the upper-right corner of the tab grid (if you Peek, you'll see it is labeled Add). You'll be taken to the standard new page screen, but notice that the number on the tab icon has increased by one. Tap the tab icon to return to the tab grid.

Long tap a tab to see a few options (9.19). From here you can:

- Close the tab.
- Close other tabs, which is useful when you only want to keep the tab you long tapped and get rid of the rest.
- Close all tabs.
- Bookmark the site in the tab you long tapped, remove its bookmark, or edit its bookmark.

When you have a bunch of tabs open and you leave Silk to do something else on the phone, Silk remembers those tabs. When you launch Silk again, you might see an alert asking if you want to restore your tabs (9.20). Silk keeps track of what tabs you had open but doesn't actively keep them open, so memory isn't wasted displaying pages you aren't looking at.

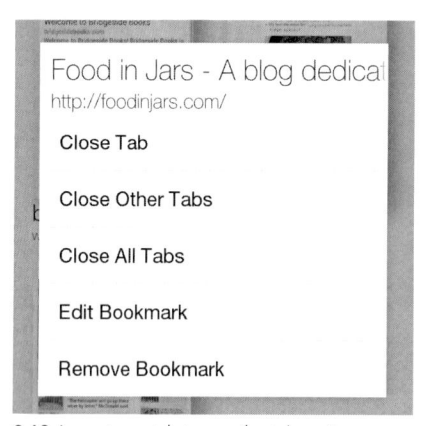

9.19 Long tap a tab to see the tab options.

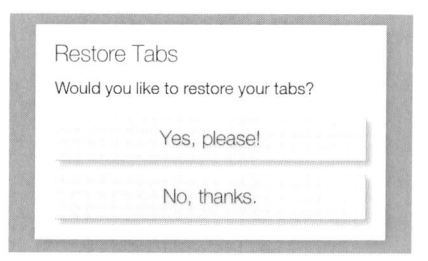

9.20 Silk keeps tabs on your tabs, so when you reopen Silk it can restore them.

Navigation

Reveal the left panel in Silk to see a bunch of options (9.21). Tap the Most Viewed drop-down menu to see the top four sites you have visited (these are the same sites that appear on the Carousel under the Silk icon). At the bottom of the panel, the Help section has some tips and tricks (though they are all covered in this chapter). Tap Send Feedback to open the Share panel. Be sure to tap Email, because the Send Feedback button, as of this writing, just generates an email addressed to silk-feedback-firephone@amazon.com, with the subject line "Silk feedback."

The other options on the left panel have a little more meat to them, so let's discuss each of them in depth.

Bookmarks

Tap Bookmarks to see your list of bookmarks, some with thumbnail previews and others with generic icons (9.22). The bookmarks are arranged in chronological order, with the site you visited most recently on top. Swipe up or down to scroll through the list. To visit a site, tap its bookmark.

Long tap a bookmark to see more options. Aside from the bottom two, they are the same options you see when you long tap a link in a webpage. This makes sense, because what are bookmarks other than a personal list of links that you frequently visit? I just blew your mind.

9.21 The left panel allows you to skip to different sections of the browser.

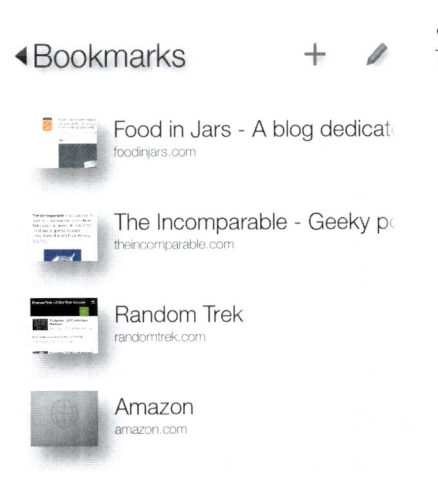

9.22 Your bookmarks. Tap a site to visit it.

The bottom two options are unique to the Bookmarks screen:

- Edit Bookmark: Tap this button to change the name and location of the selected bookmark (9.23). You can also delete it from here.
- Remove Bookmark: This deletes the site from your bookmarks. There is a confirmation alert, so you have to tap OK to confirm (or Cancel to keep the bookmark).

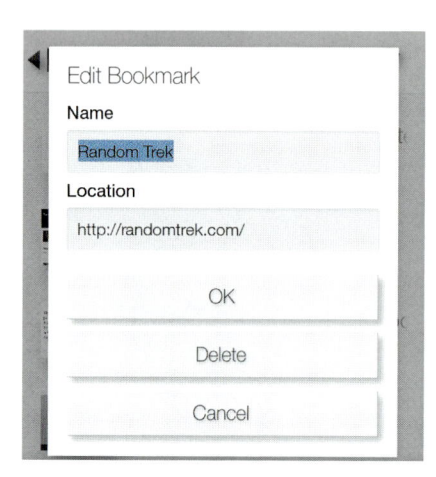

9.23 You can change the name or location of a bookmarked site.

Adding a bookmark

We've already covered several ways to add a site to your bookmarks:

- Tap the menu button at the top of any site, and tap Bookmark.
- Long tap a link, and tap Add to Bookmarks.
- Long tap a tab in the tab grid, and tap Add to Bookmarks.

You can also add a bookmark directly from the Bookmarks screen. At the top of the Bookmarks screen, tap the + icon and the Add Bookmark panel appears. Type a name (which doesn't have to be the name of the site; it can be whatever you like) and the web address. Tap OK, and the bookmark is added with a generic icon. As soon as you visit the bookmark, a thumbnail preview will appear next to it in the bookmark list.

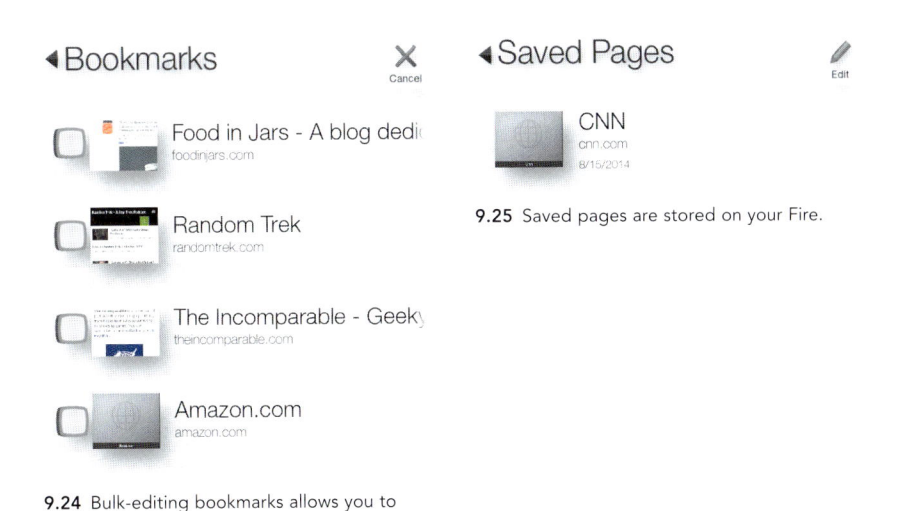

9.24 Bulk-editing bookmarks allows you to delete them all.

9.25 Saved pages are stored on your Fire.

Editing bookmarks

At the top of the Bookmarks screen, you'll see a pencil icon labeled Edit (if you Peek at it). Tap it and buttons appear next to your bookmarks, and a couple of icons appear at the bottom of the bookmarks screen (9.24).

Oddly, the only "editing" you can do here is deleting bookmarks. Tap the box next to the bookmark or bookmarks to select them. You can also tap the Select All icon at the bottom of the screen to select all the bookmarks. Then tap the Delete button. There is no confirmation screen, nor is there a way to recover deleted bookmarks (other than re-creating them from scratch).

Saved Pages

You can save webpages to your Fire's local memory. To access any pages you have saved to your phone, tap Saved Pages on the left panel. The Saved Pages screen lists all the pages currently saved on your Fire (9.25). Tap a page to open it in Silk. The only way you can tell that this page is saved on your Fire is the ⬇ icon in the top bar. Tap it to see where on the Fire the page is saved to, to copy the path to the file, or to share that location via an app on your phone.

Tapping the Edit button allows you to select all the saved pages and delete them (which will clear up room on your Fire).

Trending Now

Trending Now is one of those features that couldn't exist without the cloud.

Since your web traffic anonymously goes through Amazon's servers (unless you're on a secure site with an encrypted connection), Amazon can notice trends across Silk users. The Trending Now section lists popular sites and articles that other Silk users have been visiting (9.26). Swipe up or down to scroll through the list. Tap a page to visit it.

9.26 Trending Now shows you pages that other Fire users are visiting.

This is simply a list of links, so if you long tap any of the Trending Now entries you'll see the same options as you do when you long tap a link on a page.

Downloads

As you're browsing the web, you might come across images and files that you want to save to your Fire. Long tapping a link to a file gives you the option of downloading that file to your phone. All the files you've downloaded are listed in the Downloads section, with an icon denoting what sort of file it is (music, document, and so on) (9.27). Each entry displays the file name, the location it was downloaded from, the size of the file, and the date of download.

Tap an entry to open it in the proper app. If more than one app can open the file, you'll be able to choose which to open it with. You can opt to open it Just Once with that app, or Always (9.28).

9.27 All your web downloads are listed here.

9.28 When you tap a downloaded file, you'll need to tell the Fire which app to open it with—either just this once or always.

Long tap a file, and you can either open it or delete it. If you want to delete more than one file at once, tap the Edit button in the upper-right corner of the Downloads screen. Checkboxes appear to the left of the items, and two icons appear at the bottom: Select All and Delete. Tap checkboxes to select, or tap Select All to select all the files. Tap Delete to remove the files from your Fire.

CHAPTER 9: SILK

History

As you visit websites, Silk keeps track of them in its History panel, just in case you want to revisit a site that you failed to bookmark (9.29). Sites are grouped by the day you visited. Tap an entry in the History panel to visit it. Long tap it to see the options you'd expect from long tapping a link, with one addition: Delete. Tap Delete to remove that item from the History panel.

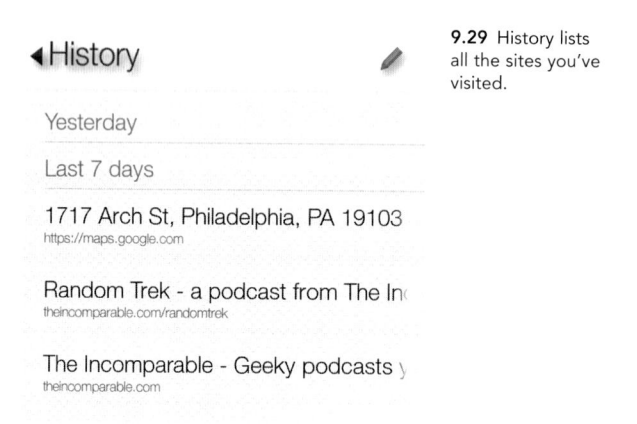

9.29 History lists all the sites you've visited.

If you have multiple entries you'd like to remove, tap the pencil icon in the upper-right corner. Checkboxes appear next to each entry, and two buttons appear at the bottom of the screen: Select All and Delete. Tap checkboxes to select entries, or tap Select All, and then tap Delete.

Swipe up from the bottom of the screen or tap the left arrow at the top of the screen to return to browsing.

Settings

You can customize your Silk experience in the Settings panel (9.30).

The first significant change you can make is which search engine Silk uses by default. Tap Search Engine to choose from Bing, Google, or Yahoo. Tap Cancel if you want to keep the setting the same.

◀ Silk Settings

Search Engine
Bing will be used for web searches.

Pop-up Windows
Pop-ups will ask before opening.

Cloud Features
Cloud Features are on.

Auto-scroll
Auto-scroll is on.

Your Data
Manage saved data, form auto-fill, cookies, location, and individual website data.

Advanced
Manage advanced settings.

Pop-up Windows
- Always
- Never
- Ask ●

Cancel

◀ Cloud Features

Cloud Features   OFF ─ ON
Accelerate page loading and enable features like Site Links.

Optional Encryption   OFF ─ ON
Encrypt traffic between this device and the Silk servers when using Cloud Features. Encryption may slow down page loads.

Instant Page Load   OFF ─ ON
Allow predictive loading to speed up page loading time.

ESV Prompt   OFF ─ ON
Prompt for Experimental Streaming Viewer.

9.30 Silk settings control your search engine, cloud features, and more.

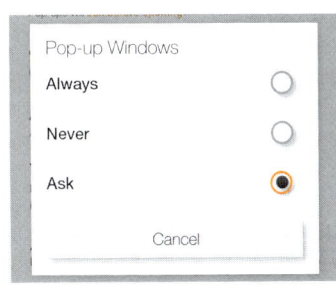

9.31 No one likes pop-ups, and your Fire can block them.

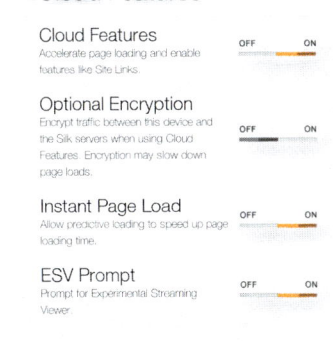

9.32 Cloud features can be turned on or off.

Next you can determine how Silk handles pop-up windows. You might not be familiar with the term, but I bet you've experienced a pop-up window. You're visiting a site and all of a sudden there's a new window with an ad in it. That new window is a pop-up. By default, Silk will ask you if you want to allow a site to create a pop-up (9.31). Tap Pop-up Windows to change this setting to Always, Never, or Ask.

Tap Cloud Features to see the following settings (9.32):

- Cloud Features: Turn off cloud features to no longer use Amazon's servers to speed up browsing. This also means that your web traffic won't be going through Amazon's servers, but rather straight to the sites that you visit. This might lead to slower browsing, but some people aren't fans of having Amazon in the middle of their web surfing. You also won't be able to use site links.

- Optional Encryption: Off by default since it can slow down web browsing, this encrypts all the traffic between your phone and Amazon's server. Having this on is more secure because no one will be able to tell what data is transferred between your phone and Amazon, but the trade-off is slightly slower loading.

- Instant Page Load: Your Fire looks at your behavior, and the behavior of other people who have visited the same page, and automatically fetches the page it thinks you're most likely to tap.

- ESV Prompt: Flash isn't supported on the Fire, but ESV (experimental streaming video) allows you to watch Flash content.

Tap the arrow at the top of the screen to go back to the rest of the Silk settings.

Auto-scroll is on by default. Tap Auto-scroll to turn it off.

The Your Data section allows you to tell Silk how to handle certain kinds of information (9.33):

- Clear Browser Data: Your browser collects a lot of data about you as you use it: the sites you've visited, your cache (files the Fire saves to make browsing faster), cookies (small bits of data from the websites you've visited), passwords, form data, and location data (9.34). Tap Clear Browser data to delete some or all of this information.

    You can selectively delete items from the list by tapping the box next to them and then tapping the Delete button. Tap Select All and then the Delete button to clear all of this data from your phone.

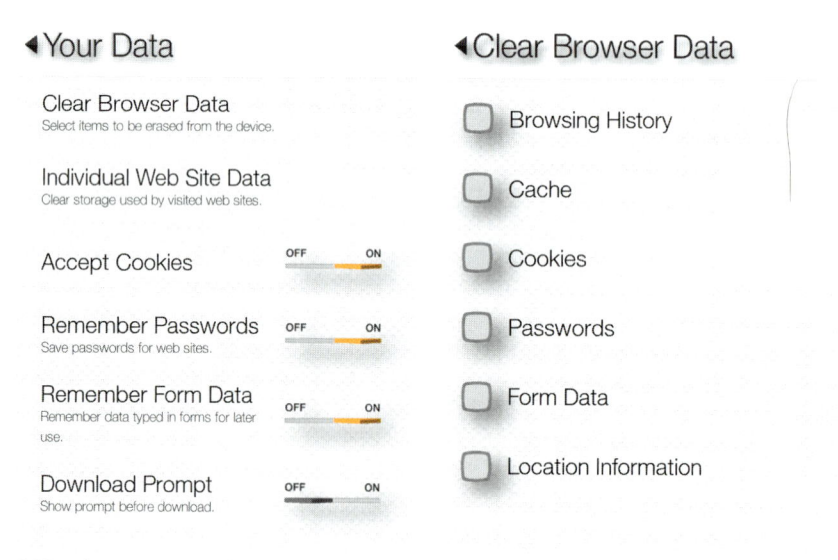

9.33 Delete your personal information.

9.34 Clear your history, cache, and more with a couple of taps.

- Individual Web Site Data: It is possible for some websites to store content on your Fire. This is fairly rare at this point, but if you tap this you can see if there is any website data on your phone. Delete it by tapping the edit icon and selecting the website.
- Accept Cookies, Remember Passwords, and Remember Form Data: These settings can be toggled on or off.
- Download Prompt: When this is on, your Fire will prompt you whenever you are about to download a file.

Tap the arrow at the top to return to the Silk settings, and turn your attention to the advanced settings, which control things like loading images, enabling Javascript, and more (9.35):

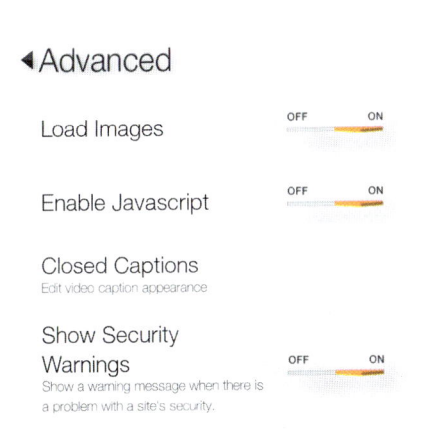

9.35 The advanced settings give you even more control over Silk.

◀ Advanced

Load Images   OFF   ON

Enable Javascript   OFF   ON

Closed Captions
Edit video caption appearance

Show Security Warnings   OFF   ON
Show a warning message when there is a problem with a site's security.

Reset all settings to default

- Load Images: By default, Silk loads images on all the sites you visit. If you want to save bandwidth (perhaps you are traveling overseas and want to avoid potential data overages), toggle this off. When you visit a page with an image, you'll be asked if you want to load images one by one.
- Enable Javascript: Javascript enables many features on modern websites, but it can cause pages to load slowly. Toggle this setting off, and you'll notice that some pages load much faster. Keep in mind, though, that many sites won't work properly without Javascript enabled.

- Closed Captions: Your Fire can display closed captions on videos. Tap Closed Captions to edit the appearance of the closed captions (9.36). You can change the font size, color, opacity, and more.

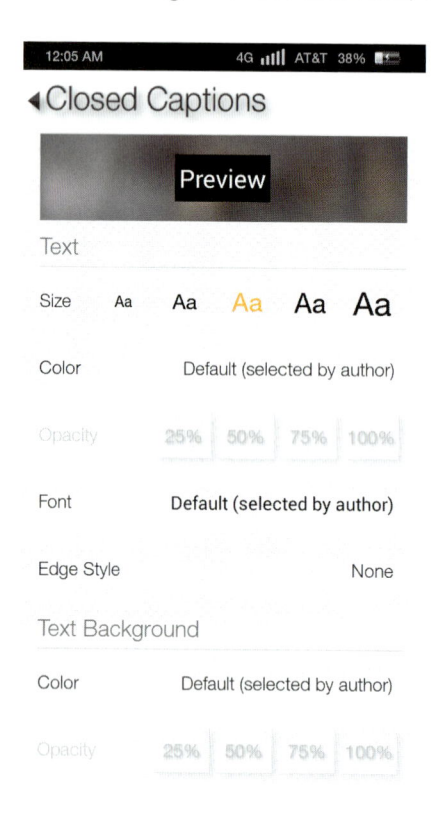

9.36 You can style the way closed captions look.

- Show Security Warnings: When a site's SSL certificate is out of date, or some other security problem exists, your Fire will warn you. If you like to live dangerously, toggle this setting off and you'll never know when you're putting your information at risk.

Finally, if you want to reset all of the advanced features to the factory defaults, just tap "Reset all settings to default." Any changes you made will be wiped away.

CHAPTER 10

Apps

Apps are what make smartphones so smart, and this chapter is dedicated to them. I cover how to buy apps from Amazon's Appstore, then I show you how to install apps from places other than Amazon's store. The rest of the chapter covers a few of the stock apps included on your Fire: Maps, Clock, and Calculator.

Amazon Appstore

The Amazon Appstore is the first place to look for Fire apps (10.1). Amazon has approved all of them, so you can be sure they won't do anything malicious to your phone. Plus, Amazon gives away a free app every day, which is a great way to dip your toe into the Appstore without spending any money.

10.1 The Appstore icon just wants you to tap it.

The Appstore is installed on your Fire, so just tap the icon and you're off (10.2). At the top of the Appstore is a rotating display of highlighted apps or categories. Tap any graphic to see the app itself or a listing of the apps in the highlighted category.

Swipe up and down to see the rest of the Appstore. There are several groupings of apps, including things like Featured Apps and Games, Free Apps & Games, and Recommended for You. The price is displayed on each app's listing, and if you Peek at them you'll see the category each app is listed in (Games, Books & Comics, and so on). Tap the group header to see all the apps in that group.

The apps

If you see an app that catches your eye, just tap to see the full listing (10.3). There's a lot of information about each app in the listing, so swipe up and down to scroll through.

The first section of the listing includes the app's icon, the name of the app in big type, the name of the developer in much smaller type, the average number of stars customers rated this app, the number of ratings, and an orange button with the price in dollars or Amazon coins (see the sidebar "Amazon Coins").

Right below the top section are screenshots and videos of the app in action so you can get a taste of what the app does. If there is a play button on the screenshot, as there is on the first one in 10.3, it is a video. Tap to play. Swipe right to left to scroll through the screenshots, and tap one to see it full size. Tap again to return to the listing.

The app description is provided by the maker of the app (as are the screenshots) to entice you to purchase the app. Tap Release Notes to see what improvements have been made to the app recently. It is always a good sign to see lots of activity in the Release Notes section, because that means the developers are actively working on improving the app.

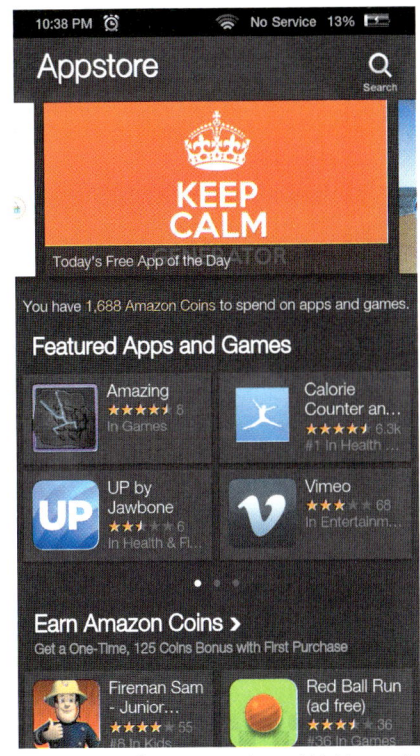

10.2 Amazon features popular and highly rated apps on the front page of the Appstore.

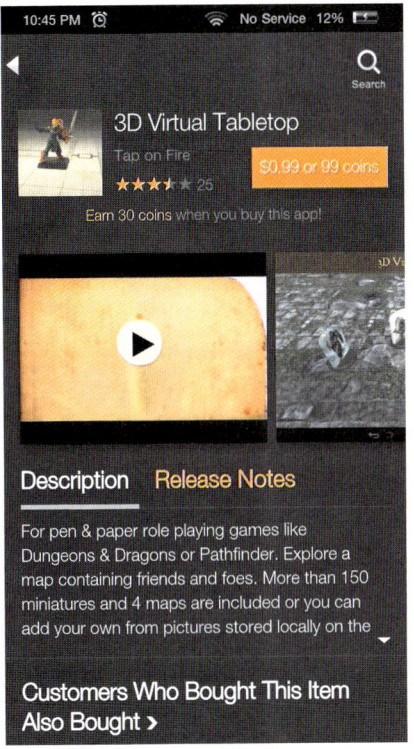

10.3 An app listing with a description and screenshots provided by the maker of the app.

Amazon is famous for their recommendation engine, and the Appstore makes use of it. Swipe left and right through Customers Who Bought This App Also Bought (10.4), or tap the header to see a list.

Amazon Customer Reviews leads off with a star chart (as I call it) that shows you how many one-, two-, three-, four-, and five-star ratings the app has received. There's a link that you can tap to see all the reviews, or you can just scroll through the selected reviews.

If you want to leave a review of your own, tap Create Your Own Review and share your opinion (10.5). There's also another opportunity to look at all the reviews: tap the See All [number] Reviews button.

CHAPTER 10: APPS 133

10.4 Reviews and recommended items are Amazon staples.

10.5 You can leave a review right from your Fire.

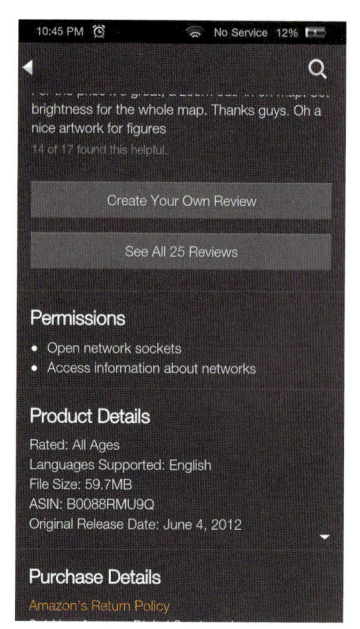

10.6 Pay close attention to the permissions that apps require.

The Permissions section lists what rights on your phone this app needs in order to run (10.6). Most apps request access to the network so they can download data or post to the Internet. A lot of apps will need permission to use the camera or to know your location. Some apps require even more access to your information: to your contacts, your phone, and more. Pay close attention to what is listed here. You want to know what you're getting yourself into when you download that free puzzle app—it might not be so free after all.

The Product Details section lists the app rating, languages supported, size of the app, and more. Tap the small white triangle to expand this field fully to see the current version number of the app.

Amazon includes a link to their return policy near the bottom of the app listing. You can also fill out a form detailing an issue you had with the app (such as if it didn't work or it tried to access something that wasn't listed in the Permissions section) by tapping the Report an Issue button. If you're really into the app, you can share a link to it by tapping Share This App. This brings up our old pal the Share panel. Select an app it would make sense to share a URL to (like email or your Twitter client), and let the world know about the sweet app you are about to buy.

Buying and installing

Once you've decided that an app is the one you've been looking for, you need to buy it. If the app is free, all you have to do is tap the orange button at the top of the listing and it will download to your Fire (assuming you have a network connection). If the app costs money, you'll need to buy it using Amazon Coins or real money **(10.7)**. Select whichever tender you'd like to use, and tap Confirm. The app downloads to your home grid.

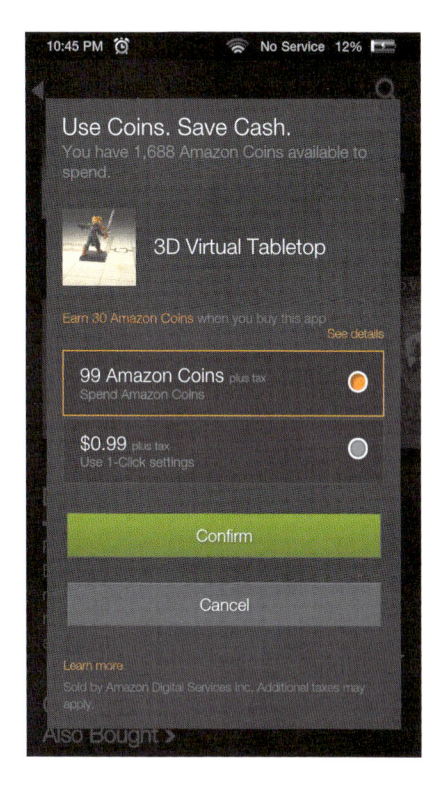

10.7 Paying for an app: Amazon Coins or real coins?

The orange bar tracks the download progress. To cancel the download, tap the X in the lower-right corner of the icon **(10.8)**. Once the app has downloaded, the install process begins automatically. As soon as it is complete, you'll get a notification. Tap the icon to use your shiny new app!

10.8 The orange bar indicates download and install progress.

CHAPTER 10: APPS   135

Amazon Coins

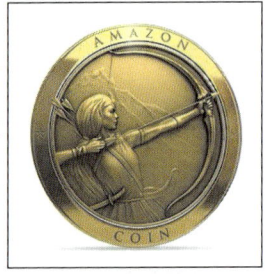

As you look around the Appstore, you'll notice that Amazon Coins keep popping up (10.9). In fact, on the front of the Appstore you're alerted to the fact that you have x number of Amazon Coins. What the heck are they?

Amazon Coins are an electronic currency only good in the Amazon Appstore. Every $1 of an app's price equals 100 coins. You will get a slight discount by paying with coins. You can earn additional coins by purchasing certain apps.

10.9 An Amazon Coin.

Why does Amazon do this? Because you need to purchase Amazon Coins in bundles of at least 500 (as of this writing, 500 coins costs $4.90), and all that money is already spent at Amazon, since you can't use them elsewhere. It is a clever way for Amazon to get your money upfront, before you even buy anything.

Are Amazon Coins worth it? I leave that up to you to figure out, based on your budget.

Test Drive before you buy

One of the biggest complaints people have about smartphone apps is that there are no demo versions. For desktop applications, it is common practice to let people try an app for a limited amount of time before they shell out any money. This lets people figure out if they actually want to buy the app, or if it isn't right for them.

Smartphone apps have traditionally not had demo versions, but the Appstore has solved this problem for some apps. Test Drive allows you to try an app without paying for it or even downloading it. No need to waste money or space on an app that isn't right for you. Sadly, not every app on the Appstore offers a Test Drive.

You'll know if an app is Test Drive eligible because there will be a Test Drive button on the app listing (10.10). Test Drive requires a Wi-Fi connection, so the Test Drive button will be grayed out when you aren't on Wi-Fi.

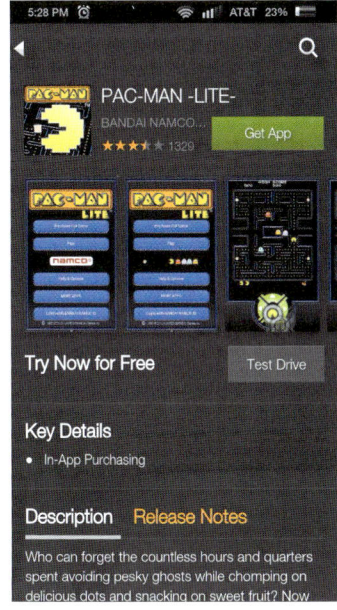

10.10 This app takes advantage of Test Drive.

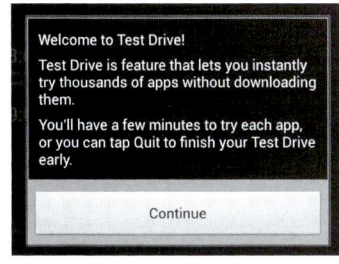

10.11 The quick explanation of Test Drive, which appears on your first launch.

10.12 Test Driving Pacman.

When you tap a Test Drive button for the first time, you'll get an alert explaining what Test Drive is and then it will set up the Test Drive session (**10.11**). Keep in mind that the Test Drive is timed (10 minutes, generally), so watch the clock in the upper-left corner of the Test Drive window.

What's happening here? All of the app's functionality is available, but the app is running on a server somewhere in Amazon's data center. The app is streamed to your phone so you can interact with it, test out the interface, try the features, and decide if you want to buy it (**10.12**). To purchase it, tap the orange button in the upper-right corner, which displays the price. You'll be taken back to the app listing and asked to confirm your purchase.

▶ **NOTE** Test Drive is heavily dependent on a fast Wi-Fi connection. If your network slows down, you'll get an error message telling you that you can no longer Test Drive with your connection.

CHAPTER 10: APPS   137

In-app purchases

You can make in-app purchases of additional functionality (or, say, more lives in games). Some developers create apps that are free to download and then include a number of features, or digital items, for sale in the app itself.

An app's listing will tell you whether it includes in-app purchases. How in-app purchasing looks in each app is dependent on the makers of the app. Each design is different, but the result is the same: you pay real money for something digital.

To complete an in-app purchase, you're taken to a familiar screen that shows you the price (in both dollars and Amazon Coins) (10.13). Tap whichever mode of payment you want to use, and then tap Get Item.

10.13 An in-app purchase.

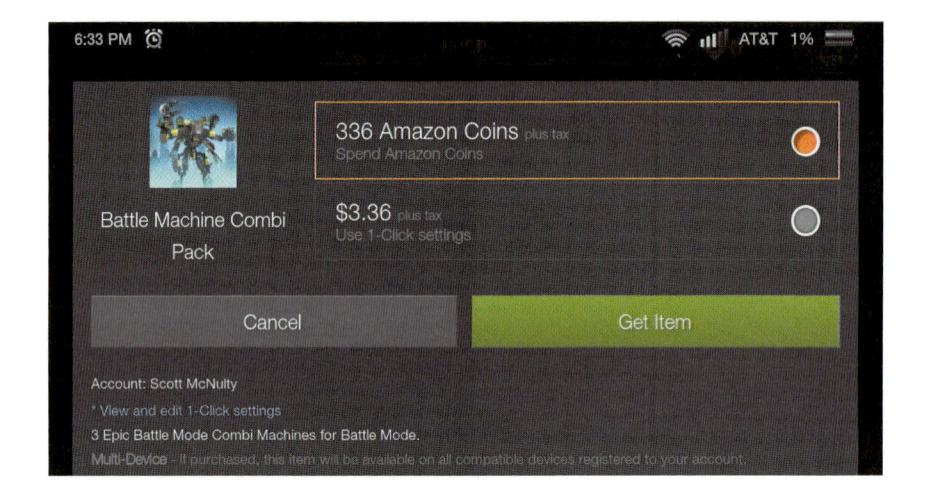

The first time you purchase something in-app, you'll have to provide your Amazon account password and you'll be asked if you want to require the password for future purchases. Once you enter that information, you're taken back to the app. Tap Cancel if you change your mind.

Buying Amazon Coins

On the front page of the Appstore, you'll see the number of Amazon Coins you have in your account. Tap the number to buy even more coins (10.14). How nice!

10.14 You can buy Amazon Coins on your Fire.

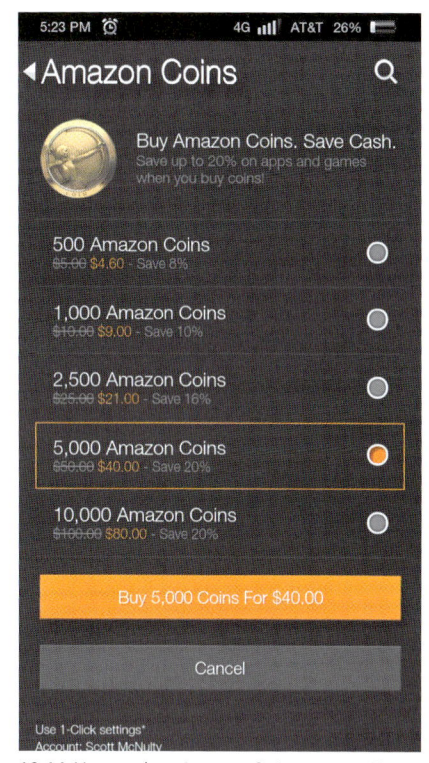

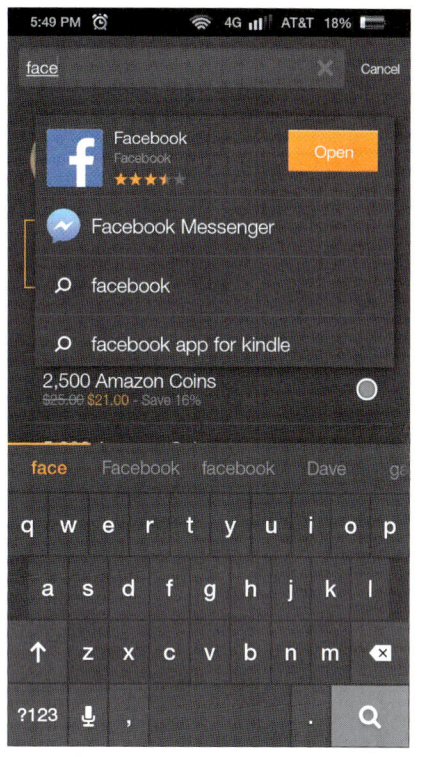

10.15 The Appstore search is pretty intelligent.

Navigating the Appstore

There are a number of ways to find apps in the Appstore. No matter where you are in the Appstore, the search icon (a magnifying glass) appears in the upper-right corner. Tap it, and type your search query in the search field (10.15). As you type, your Fire will suggest possible search terms (displayed with an magnifying glass icon to their left) and put a couple of apps in the top of the search results. The top suggestion is the app your Fire is pretty sure you meant. Tap the orange button to install that app (or open it if you already have it installed) without going into the app listing.

You can also tap either the search button on the keyboard or one of the suggested searches to see the full list of results (10.16). Each result has the app's icon, name, rating, and price. Bring out the right panel to see a list of additional suggested search terms (10.17).

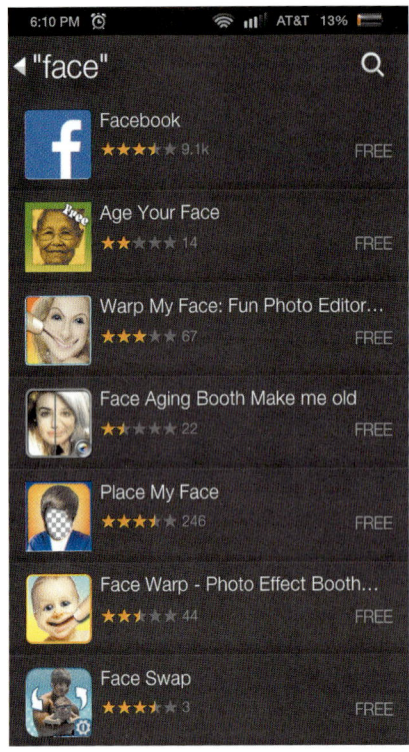

10.16 Search results include ratings and app prices. Peek to see the app categories.

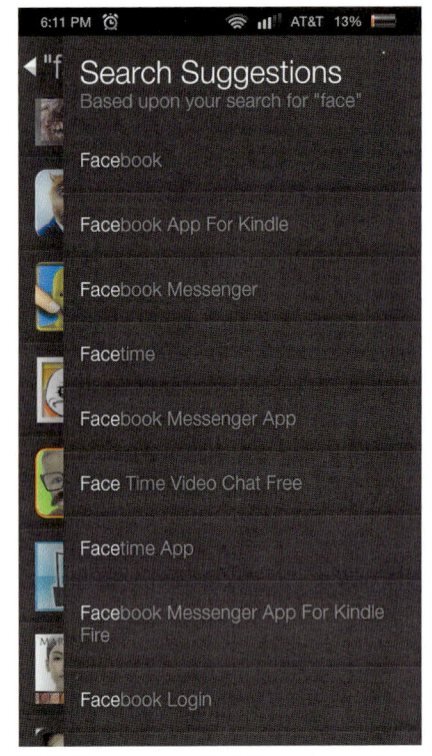

10.17 The search results right panel suggests further search terms.

Left panel

If searching isn't for you, the left panel is a great place to hop to different categories of apps in the Appstore (10.18). Right at the top is a link to your apps. Tap it, and you're taken to the home grid. Your Amazon Coin balance is displayed; tapping it gives you the option to purchase more coins.

Swipe up to scroll through the complete category listings. Some highlights include:

- Recommended For You: These apps are based on the apps you already own.
- Best Sellers
- New Releases
- Test Drive: All the apps listed here offer a Test Drive.

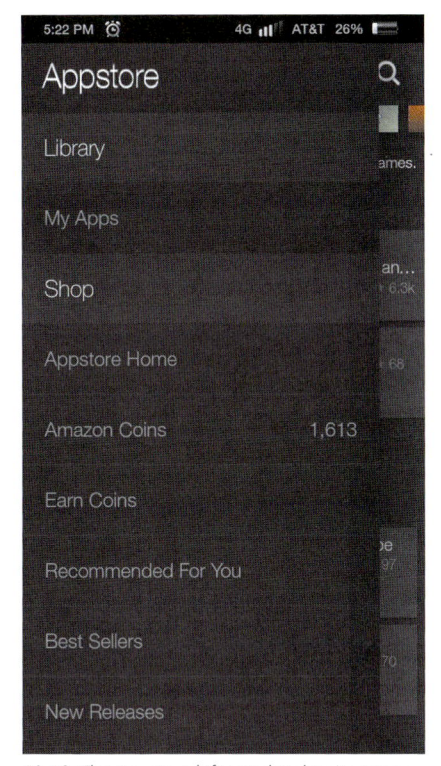

10.18 The Appstore left panel makes it easy to go to different sections of the Appstore.

10.19 All the categories of the Appstore.

Below those "meta" categories are the actual categories in the Appstore. These include Games and Entertainment. Tap View All Categories to see the entire list (10.19).

Back on the left panel, the More section has a few interesting options. App Updates lists any apps that have recently been updated, along with the changes included with those updates (10.20). By default, your Fire will automatically update apps in the background, but you can change that setting.

You can find periodicals, like newspapers and magazines, in the Appstore. Most of the apps are free but offer subscriptions as an in-app purchase. Tap Subscriptions to manage them (10.21). Here you can see when your next payment is due, and toggle on and off auto-renewal (10.22). You can also opt out of sharing contact details with the publisher here.

10.20 App Updates happen automatically, but the changes are recorded here.

10.21 Managing your subscriptions in the Appstore.

10.22 Here you can turn off auto-renew and change your privacy settings on a per-subscription basis.

Sideloading

Another method of getting apps onto your Fire is little more involved than just tapping a couple of buttons. The Fire, by default, only installs apps that are available from the Amazon Appstore, but there are a number of other Android app stores (including the Google Play Store, which is the standard app store for most other Android devices).

> **NOTE** Keep in mind that although your Fire runs an operating system called Fire OS, it is, underneath all the Amazon trappings, an Android phone.

It is possible to install, or *sideload*, apps from other app stores onto your Fire. Why would you want to do this? As I mentioned, any app in the Amazon Appstore is there because Amazon approved it. This means that some apps aren't in the store because Amazon rejected them, or because the developer didn't want to list them in the Amazon Appstore for whatever reason. This doesn't mean you can't install the app on your Fire; it just requires more work.

First, you need to enable sideloading. Go to Quick Settings > Settings > Applications & Parental Controls > Allow Non-Amazon App Installation. Toggle App Installation from off (the default) to on. A warning pops up letting you know that Amazon takes no responsibility for any damage you may do to your Fire with these crazy apps (10.23). Tap OK, and now you can sideload all the apps you want.

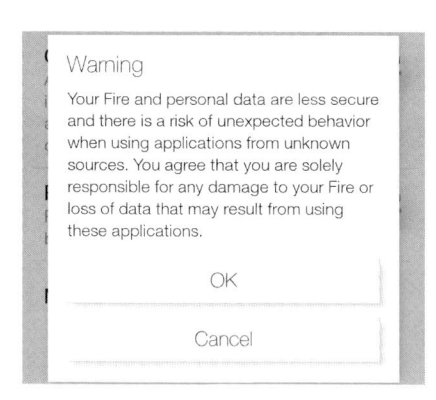

10.23 The dire warning about installing non-Amazon Appstore apps on your Fire is a little scary, but nothing will happen to your Fire. Probably.

Every app on your Fire is just a collection of files grouped into an .apk file. To sideload an app, you need to first find the app's .apk file. Some Android app developers allow you to directly download the .apk file from their website (Firefox does), though most don't. The easiest way to get your hands on the .apk file for a particular app is to install it on another Android device that can install it, and then back up the file onto your computer or another form of removable media using a backup app.

Once you have the .apk file, you can do one of two things: email the .apk file to your Fire or transfer it using USB.

Emailing the .apk is simple. Once the email is on your Fire, tap the attachment to download it. Tap again, and the install process begins with a list of permissions the app needs to run (10.24). Tap Next, and the install process starts (10.25).

10.24 Before a sideloaded app installs, it tells you what permissions it needs.

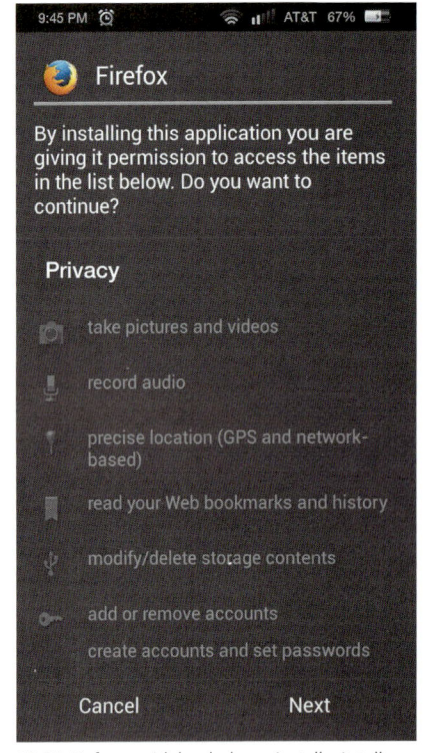

10.25 The installation process happens speedily.

Transferring via USB on a Windows computer

1. Connect your Fire to your computer with a micro-USB cable.

   The Fire appears on your computer as an external drive.

2. Double-click the external drive icon that represents your Fire under My Computer, create a new folder, and name it Custom Apps.

3. Drag and drop the .apk files into the Custom Apps folder.

4. Disconnect your Fire from your computer. On the Fire, go to the Amazon Appstore.

5. Download a file browser app. I like File Manager by Rhythm Software, but there are a few options available on the Appstore.

   File browser apps allow you to look at hidden folders and directories, like the one you just created, on your Fire. Use that app to locate the directory with the .apk files in it.

6. Tap the .apk file.

   The app installs and adds itself to your home grid.

Transferring via USB on a Mac

1. Download the Android File Transfer app for your Mac from www.android.com/filetransfer/.
2. Connect your Fire to your Mac using a micro-USB cable.
3. Launch the Android File Transfer app on your Mac, and it will present you with the full directory list (10.26).
4. Create a new folder by clicking File > New Folder, and name it Custom Apps.
5. Drag and drop the .apk files into the Custom Apps folder.
6. Disconnect your Fire from your Mac. On the Fire, go to the Amazon Appstore.
7. Download a file browser app, such as File Manager, from the Appstore.

   File browser apps allow you to look at hidden folders and directories, like the one you just created, on your Fire. Use that app to locate the directory with the .apk files in it.
8. Tap the .apk file.

   The app installs and adds itself to your home grid.

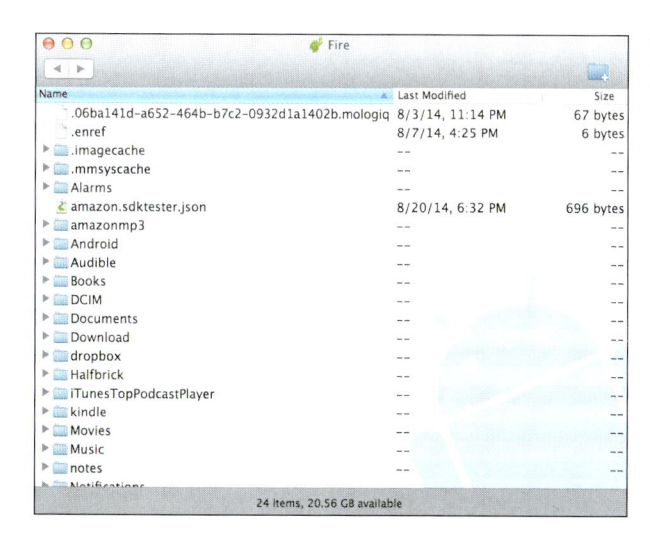

10.26 The Android File Transfer app for the Mac.

CHAPTER 10: APPS

Keep in mind that although your Fire runs the Android operating system, it is a heavily modified version of it. Apps from other app stores won't be optimized for your Fire, so they might not work. Or they might crash or stop working at any moment. Chances are slim that you'll do any damage to your Fire by sideloading an app, but the app you sideload might not work as intended.

Appstore settings

Open the left panel in the Appstore, swipe up until you see the More section, and tap Settings **(10.27)**. In-app purchases are enabled by default. If you want to disable them, tap In-App Purchasing and uncheck the box. Now in-app purchases will not be available in apps, though you can always turn them back on.

Your Fire automatically updates your apps in the background. If you want to change this behavior, tap Automatic Updates and you'll see three options **(10.28)**:

- Enable Automatic Updates

10.27 The Appstore settings allow you to turn off in-app purchasing and auto updates.

10.28 Fine-tune how, and if, your apps automatically updated.

- Use Cellular Data is disabled by default, so your apps only update when you're connected to a Wi-Fi network. Tap the box to enable updating over cell data, but keep in mind that your data plan might have a monthly limit, and apps can be large.

- Notify Me When Updates Are Installed is enabled by default. When an app is updated, you get a notification in Quick Settings. Don't want to know? Uncheck this box.

External Market Links determines how your Fire behaves when you tap a link to an app on an app store other than Amazon's (10.29). By default, your Fire will warn you that this link is to a non-Amazon app store and asks if you'd like to try opening it in the Amazon Appstore. If this isn't possible, you're just taken to the front page of the Appstore.

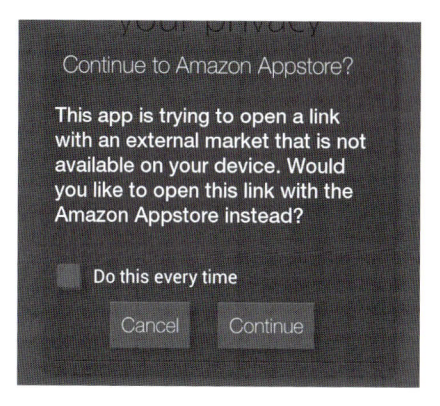

10.29 Tap a non-Amazon Appstore app link and you'll get this message.

Stock Apps

A host of apps come preinstalled on your phone. This section covers using some of the standard apps to get things done—and in some cases, to get where you're going!

Games

The Games app gathers the games you already own into one place (10.30). Tap a game to download it from your Cloud Library (if it isn't already installed on your Fire) and install it. Launch and enjoy!

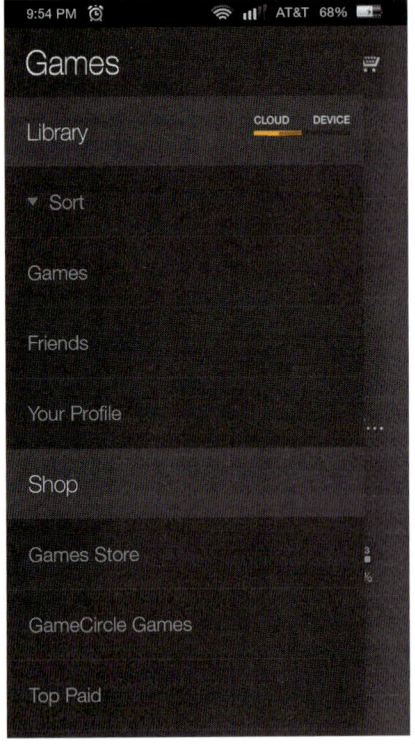

10.30 The Games app isn't really an app. It is a collection of the games you already own.

10.31 The Games left panel helps you find more games, and friends who play them.

The left panel Library section allows you to sort this list differently (by name or by most recently played) and decide which Library you want to display (Cloud or Device) (10.31).

The Friends and Profile options are part of Amazon's GameCircle service, which allows you to find other people who are playing the same games as you and compare scores or play against each other. Tapping Friends lists your friends and allows you to search for new friends. Tap Profile to create a profile name and select an avatar, or image, that is associated with your GameCircle profile (10.32). The Shop section of the left panel links to parts of the Appstore that are related to games. If you're interested in trying out a GameCircle game, tap that link to be taken to the GameCircle page on the Appstore.

Turn off your GameCircle profile by tapping Settings and toggling Share Your GameCircle Profile to off (10.33). People won't be able to find you on GameCircle, though you can still track your own accomplishments.

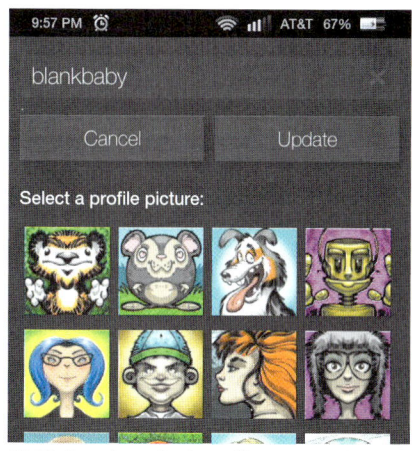

10.32 Your GameCircle profile consists of a nickname and a profile picture.

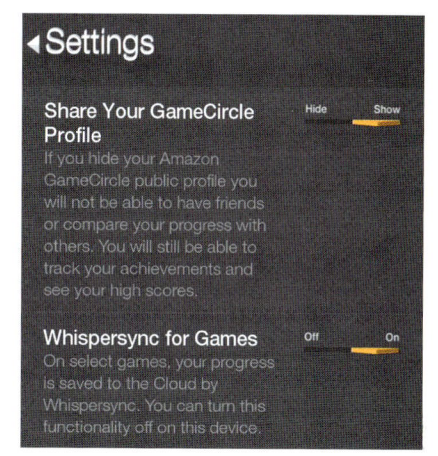

10.33 Toggle Share Your GameCircle Profile to Hide if you don't want people to see your accomplishments.

You can also toggle off Whispersync for Games. Whispersync, for games that support it, saves your progress to Amazon's server. With Whispersync on, you can play a supported game on your Fire phone, quit it, and then launch that same game on your Fire HDX tablet and pick up right where you left off.

Maps

The Maps app shows you where you are and gives you directions. Maps offers turn-by-turn navigation for any of the direction types.

Launch the Maps app, and by default it displays your current location (10.34). Notice that an arrow appears in the status bar. This means that an app (in this case, Maps) is using your phone's location.

Swipe to move the map around and explore your surroundings. You can also pinch in to zoom in to the map, and pinch out to zoom out. You can always return to your current location by tapping the icon.

At the top of the screen is a search field. Enter either an address or a place or business name. As you type, your contacts and recent locations show up as suggestions. Entering a place name—say, McDonald's—displays points on the map that match that search (10.35). Tap the list icon next to the search bar to display a list of all the results (10.36).

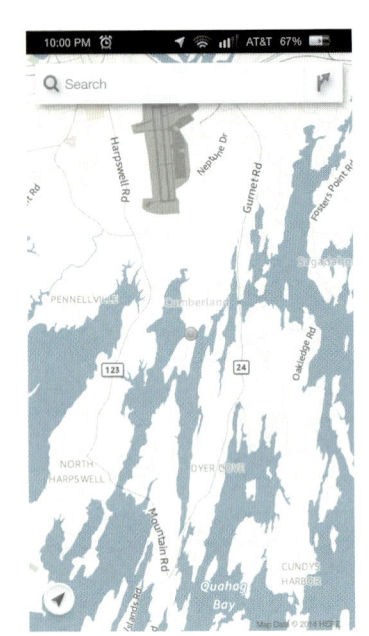

10.34 My current location displayed in the Maps app.

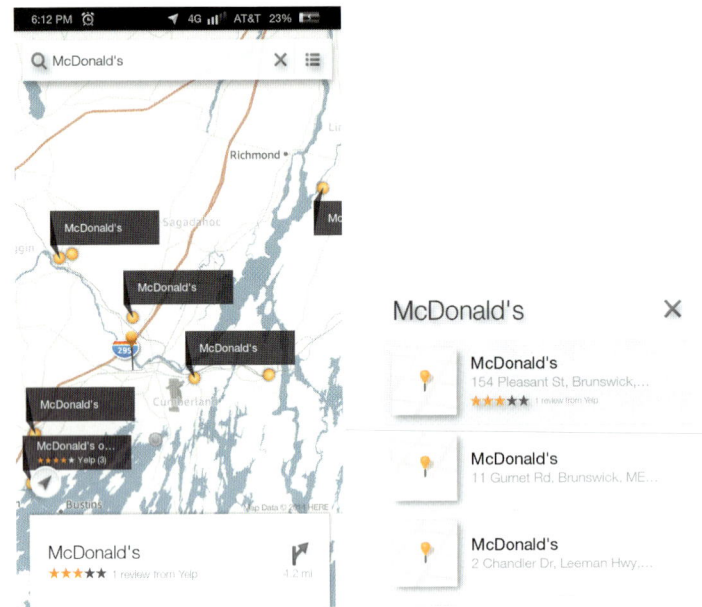

10.35 Search results in the Maps app are displayed as points on the map. Here we're using Peek to see more information about each point.

10.36 The list of search results.

Tap a point on the map to bring up a card with information about that location and the distance from where you are. Tap the card to see all the information listed (tapping an entry in the list view takes you right to the full card) **(10.37)**. From this card, you can share the location, bookmark it for quick future navigation, or create a new contact with this information. Swipe left and right to cycle through the cards for all the search results.

▶ **TIP** Long tap a point on the map to drop a pin. You'll get the same options as you do when tapping a search result: Share, Bookmark, and New Contact.

Tap the distance to the location and select what type of directions you'd like: walking, transit, or driving **(10.38)**. You can even change the starting point by tapping it to get directions from somewhere other than where you are. Tap Get Directions, and the route is created. You're shown a preview with the route, the distance, the length of the trip, and an estimated time of arrival **(10.39)**. Tap the list icon to see the full directions.

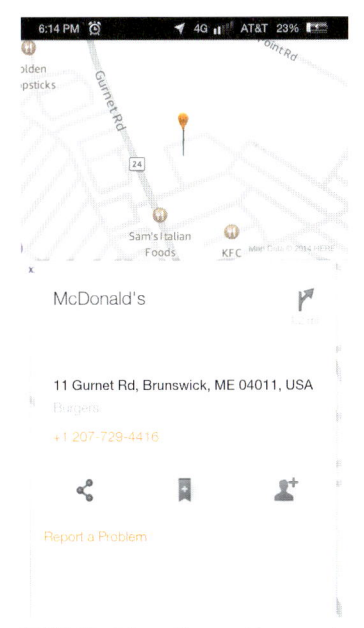

10.37 The information card for one of the points on the map.

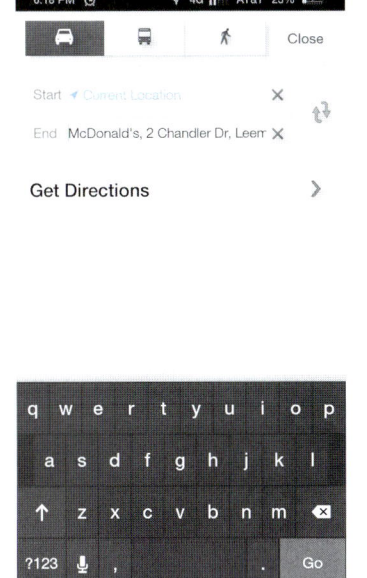

10.38 You need to tell Maps what kind of directions you're in the mood for.

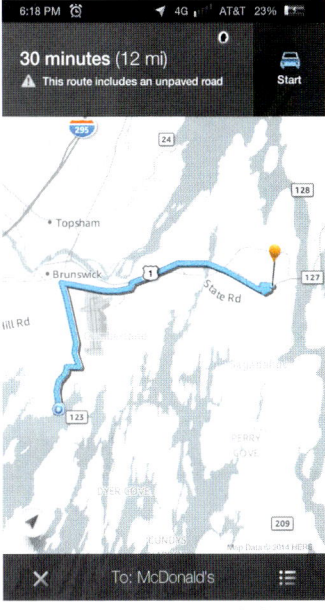

10.39 The route is previewed, along with the estimated length of travel.

Tap Start and your Fire will give you audible turn-by-turn directions (10.40). At any point during the route, tap the screen to pause the directions and bring up the controls to increase the voice volume, see the list of directions, or cancel the route entirely.

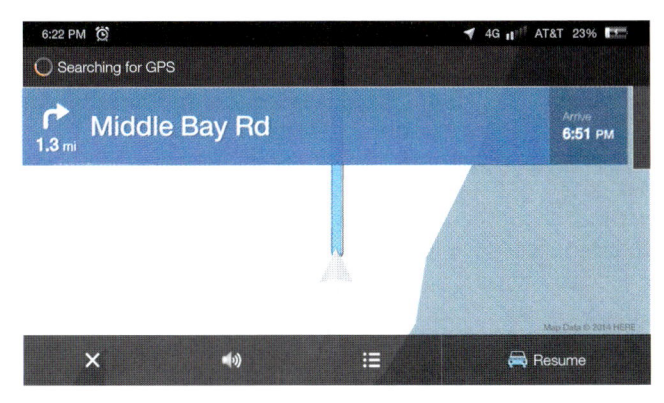

10.40 As you travel, your progress is shown by an arrow that represents you on the map.

CHAPTER 10: APPS   151

The left panel controls the map layers. By default you see a stylized map; tap Satellite or Traffic to add additional data (green means no traffic, yellow means congested, and red is not good) (10.41). You can tap any of the construction or delay icons on the traffic layer to get more information about them (10.42).

▶ **TIP** Some cities' maps include 3D buildings that you can see at different angles with Peek.

The right panel lists your bookmarked locations as well as your recent searches. Tap one to see it on the map.

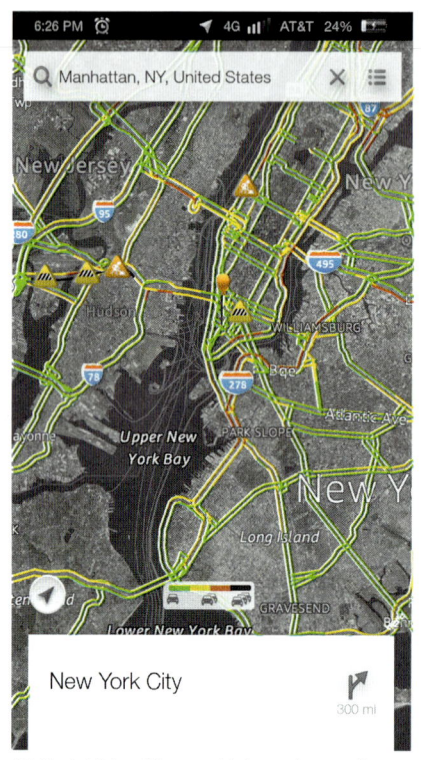

10.41 Additional layers of information on the map include satellite imagery and traffic data.

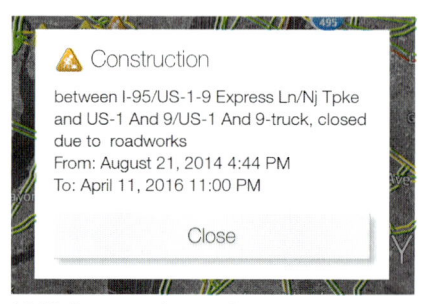

10.42 A construction warning.

Clock

The Clock app has four distinct functions: alarm clock, world clock, timer, and a stopwatch. Tap the clock icon in the home grid to launch the clock. Your current alarms are displayed, along with icons on the bottom to switch to the different parts of the clock (10.43). You can have as many alarms as you like, and each can be toggled on and off separately.

Tap the + to add a new alarm (10.44). Set the time you want the alarm to go off by swiping the hour, minute, and PM/AM. Tap Repeat to set which day of the week this alarm should always go off on (by default new alarms don't repeat), and then tap the Sound section to pick an alert sound. Tap Save and you have a new alarm. An alarm icon appears in the status bar . When the alarm goes off, an alert will be displayed on your screen and your selected sound will play. Tap Off to silence the alarm, or tap Snooze to silence it for a specified amount of time.

Tap the world clock icon to see times from around the world (and your current location) (10.45). Tap the + to add a city to the list. Tap a suggestion and it is added to the list. Long tap an entry to remove it.

10.43 The Clock app can store multiple alarms, each of which can be turned on and off independently.

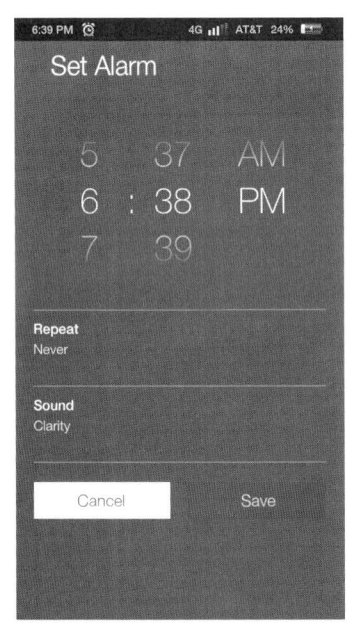

10.44 Setting an alarm.

10.45 The world clock can be set to display the time in any city you're interested in.

CHAPTER 10: APPS   153

The timer ⏲ plays an alert when a set amount of time has passed. As the time is counting down, an Add 1 Minute button appears, so you can adjust the time on the fly, along with Pause and Cancel buttons (10.46). You can only have one timer running at a time.

Stopwatch ⏱ lets you time things using your Fire (10.47). Laps are supported, so you can time races and the like.

Tapping the clock's Settings button at the top right everywhere but in the Stopwatch screen brings up a few options (10.48). Here you can change the date and time on the device (by default it is automatically configured), change the volume just for the clock alerts (so you can have your phone set to silent for most things but still have your alarm clock wake you up), and toggle on and off vibrating.

Tap Snooze Length to set the length of the snooze button feature—anywhere from 1 to 30 minutes (it is 10 by default).

10.46 The timer is a timer. When it is running, the Add 1 Minute button appears.

10.47 The stopwatch supports Laps.

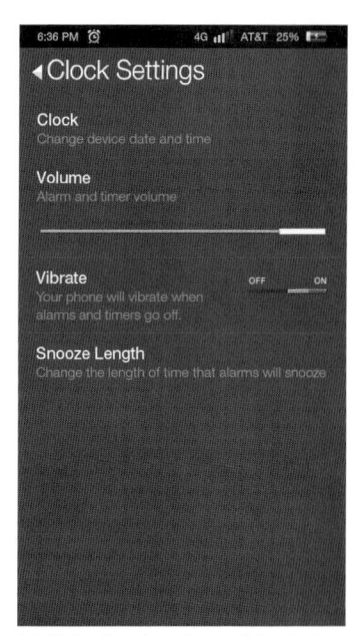

10.48 Setting the volume of the clock alarms independently of the system sounds means you can silence your phone but not sleep through your alarm.

154   THE AMAZON FIRE PHONE

Calculator

The Calculator (10.49) adds, subtracts, divides, and multiplies with the best of them. But if you rotate your Fire into landscape mode, it becomes a scientific calculator with a host of advanced features that I haven't used since the last time I was in a math class (10.50).

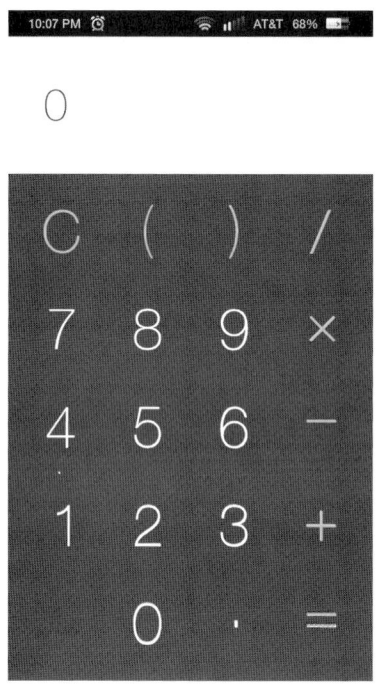

10.50 Turn the calculator on its side, and more options are revealed.

10.49 The simple calculator.

CHAPTER 10: APPS   155

My favorite feature, though, is in the right panel (10.51). The tip calculator will tell you exactly how much tip to leave based on the amount of the bill, the percentage you want to leave, and how many people are in your party.

10.51 The tip calculator will save many a friendship.

CHAPTER 11

Camera and Firefly

Your Fire is equipped with two cameras: a 13-megapixel camera on the back and a front-facing camera that you can use for video conferencing, selfies, and more. This chapter covers taking pictures with either camera, using the camera's features, viewing your pictures, editing basics, and using the camera's most innovative feature: Firefly.

You can quickly start the Camera app on your Fire by pressing the shutter button on the left side of the Fire (right under the volume buttons) (11.1). The majority of the interface is dedicated to a viewfinder showing you what you're taking a picture of, but there are a few buttons to explain:

11.1 The Camera app is almost entirely a viewfinder.

- **Shutter** ⦿ Tap this button to take a picture. You can also use the physical shutter button to take a picture.
- **Still/Video** 📷 The camera is capable of taking still pictures and video. Tap this button to switch modes.
- **Flash** ⚡ There are three flash settings, which you can cycle through by tapping the flash icon: off, on (which will use the flash no matter the lighting situation), and auto (which will fire the flash only when it is needed).
- **Camera selector** ⇆ Tap this button to switch between the front-facing camera (for taking selfies) and the camera on the back of the phone (for pictures of things other than yourself).
- **Your photos** 🖼 Tap this icon to look at the photos on your camera roll.
- **Settings** ⚙ Your camera has a few settings and modes that can be engaged using this icon.

Taking a Picture

To take a photo with your Fire:

1. Launch the camera app by using the shutter button on the side of the phone or by tapping the icon in the home grid.
2. If you want to take a picture in landscape (the photo will be wider than it is tall) turn your Fire on its side. The icons will swivel to point in the correct direction. For portrait-oriented pictures hold your Fire upright.
3. Tap the screen to focus on different things in the viewfinder. A square will appear where you tapped to let you see where the focus is (11.2).
4. Press the shutter button (physical or on the camera app) and a picture is taken.
5. Review the picture (11.3). If you like the picture tap the checkmark; if you want to discard it tap the X. You're taken right back to the camera to take another picture once you're done.

▶ **TIP** To capture a series of pictures in rapid succession hold down the shutter button icon (this won't work with the physical button) and you'll take a burst of pictures with the number of pictures displayed next to the Your photos icon on the camera. This is handy when the subject of your pictures isn't standing still.

11.2 The focus brackets appear when you tap the screen to shift the focus.

11.3 Tap the X to toss the photo and the check to save it.

Taking a panorama

Sometimes you want to capture an image of a sweeping vista, or an entire room, that just won't fit in one picture. That's where the panorama feature of the camera comes in. It allows you to take lots of pictures and then automatically stitches them together into one image. Here's how:

1. Launch the Camera app, and tap the settings icon (11.4).

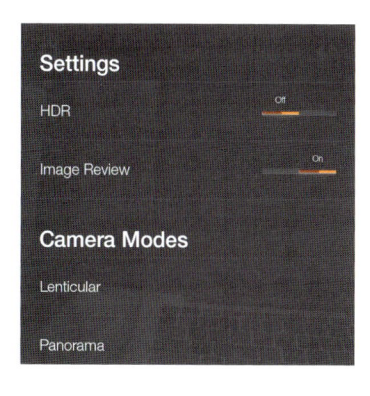

11.4 The Camera settings give you access to the different modes.

CHAPTER 11: CAMERA AND FIREFLY 159

2. Tap Panorama. Instructions and arrows indicating which way to move the phone appear onscreen.

3. Tap the shutter button icon and move your phone up/down or left/right, making sure to keep the yellow arrow on the line in the guide (11.5). Don't move it too quickly, or your Fire will show a "Too fast" error. Also, move the phone in only one direction. Trying to move it up and left confuses the app.

11.5 Panorama mode lets you take pictures of sweeping vistas, or hotel rooms.

4. Tap the shutter button when you've captured what you're after, and the photo is saved.

Lenticular photography

Dynamic Perspective gives the Fire's interface a little depth, but wouldn't it be cool if you could take your own pictures and capture some of that same magic? Using the lenticular camera option makes this possible. This option takes as many as 11 separate photos and sandwiches them together. When you're looking at one of these photos on your Fire and move it around you'll see different photos depending on the position of your head. Here's how to do it:

1. Launch the Camera and tap settings.

2. Tap Lenticular and some instructions appear (11.6).

3. Tap the shutter icon to take a picture. The grey box under the shutter button shows you the number of pictures you've taken (it has to be somewhere between 2 and 11), and all the pictures you've taken will be lined up to the left this box. Swipe to select the previous images. If you tap the X while looking at a previous image it'll be removed from the lenticular picture.

   As you're shooting, the previous picture will be ghosted on the screen so you can situate the next picture correctly (11.7).

4. Once you're happy with the pictures tap the arrow pointing to the right, and your Fire will, in a couple of seconds, create the image and then preview it for you. Tilt your Fire to see the lenticular action. Tap the check box and it is saved. Tap the left arrow to return to the camera and take more pictures for your lenticular picture.

11.6 Lenticular mode takes multiple images and makes them one.

11.7 Ghosting helps you plan your next shot.

CHAPTER 11: CAMERA AND FIREFLY

HDR

HDR (high dynamic range) is a method of photography that boosts colors and handles backlighting a little better than normal photography. In fact, when you're taking a picture of a subject against a source of light, the camera will suggest you turn on HDR (11.8). Tap the HDR Recommended alert to turn on HDR for that photo, and then take the photo as usual.

If you want to specifically take an HDR photo:

1. Launch the Camera app and tap the Settings icon.
2. Toggle HDR on, and an HDR icon will appear in the viewfinder (it won't appear on your photo) (11.9).
3. Tap or press the shutter button, and you've captured your HDR photo!

Tap the HDR button in the viewfinder to take the camera out of HDR mode.

11.8 This alert appears when you're taking a picture of a backlight subject.

11.9 Tap to turn off HDR.

Shooting a Video

1. Launch the camera app, and tap the Camera selection icon to switch to video mode.
2. Decide if you want your video to be in landscape mode (which is the common way to do video) or portrait. Hold your Fire accordingly and tap the record button. The recording begins (11.10).
3. As you record a timer appears in the upper left corner letting you know how long you've been recording.
4. Tap the Record button one more time to stop recording and save the video.

11.10 Recording a video.

Pictures

As you take pictures, the camera saves them to your camera roll. You can open the camera roll in two ways: Tap the pictures icon in the camera and that will take you directly to the camera roll, or tap the Pictures app in the home grid.

Either way you're taken to the same app: the Photos app (11.11). As you can see in the figure, the photos are displayed in a grid. Peek, by tilting your phone, to see when the pictures were taken or saved. Swipe up and down to scroll through all the pictures on your camera roll.

The camera roll isn't the only album of photos on your Fire. While looking at the photo grid, bring out the left panel and you'll see the other albums on your Fire (11.12). The checkmark indicates which album you're currently displaying. At the top, All lists all the photos on your Fire and on your Amazon cloud storage (including ones you took and ones you saves from the Internet), Videos show just videos, Device displays photos and videos on your Fire, and Cloud Drive shows you photos and videos on your Cloud Drive. Tap one to switch to that view.

CHAPTER 11: CAMERA AND FIREFLY   163

11.11 The Photos app shows all your pictures in a grid.

11.12 The left panel lets you decide which album you want to view.

11.13 When you view a photo you have four options at the bottom: take another picture, share this one, delete it, or edit.

Tap a photo or video (indicated by a play button) and it'll be displayed (11.13). Along the bottom of the screen (while you're looking at a photo) you will see four icons:

- **Camera.** Tap this to open the camera app.
- **Share.** This will bring up the sharing panel so you can send this photo to another app (to email it, for example, or text it to someone).
- **Delete.** Don't like the picture? Delete it by tapping this icon and then tapping Delete in the popup menu (11.14).

11.14 Tap Delete to get rid of a picture from your Fire.

- **Edit.** You can do some light photo editing on your Fire. See the next section for details.

Long tapping a photo brings up the option to share or delete that photo. If you want to delete or share multiple photos, go to the grid, tap the select icon at the top right of the screen, and tap the photos to select them (11.15). Tap the share or delete button at the bottom of the screen to complete the action.

The right panel displays information about the photo, including when it was taken, its location on your Fire (or in the Cloud Drive), and its dimensions.

▶ **TIP** Videos only display the Share and Delete buttons.

To play a video, tap the play icon (11.16).

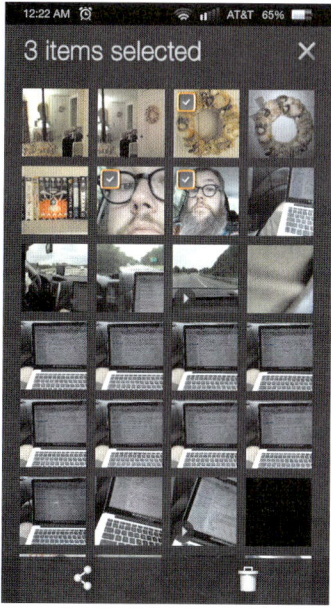

11.15 Select multiple photos to share or delete.

11.16 The play icon indicates that this is a video.

CHAPTER 11: CAMERA AND FIREFLY

Firefly

Firefly turns your Fire's camera into a scanner of real world items. Point it at a book, for example, and it'll return the price of that book and a link to buy it on Amazon. Point it at a sign with a phone number and URL and Firefly will pull out that information and make it tappable so you can call that number without having to type it. Firefly isn't limited to text and physical objects, however: It can identify movies, TV shows, and music as well. When it works, which is most of the time, it really feels like magic.

▶ **NOTE** Firefly requires a network connection.

To start Firefly either tap the Firefly icon in your home grid or hold the shutter button on the left side of the Fire for a couple of seconds. Firefly launches and looks very much like the Camera app, with some white specks floating about. Those are the fireflies, and they swarm toward any information or product that they recognize.

When you point Firefly at a book, for example, the fireflies swarm all over it and it is identified (11.17). Tap the identification to see more information, buy the book on Amazon, or share this with someone else (11.18).

▶ **TIP** Pointing Firefly at an item's barcode greatly increases the chances that the item will be recognized successfully.

Pointing Firefly at a URL will highlight the URL and then display it at the bottom of the screen as a link (11.19). Tap the link and you're taken to Silk, which loads the page. The same process happens for email addresses and phone numbers (11.20).

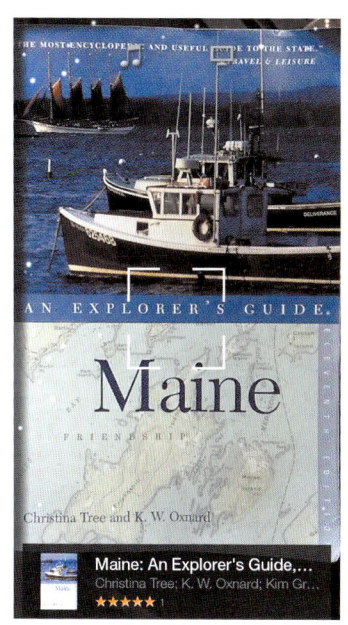

11.17 Firefly easily identifies books based on their cover.

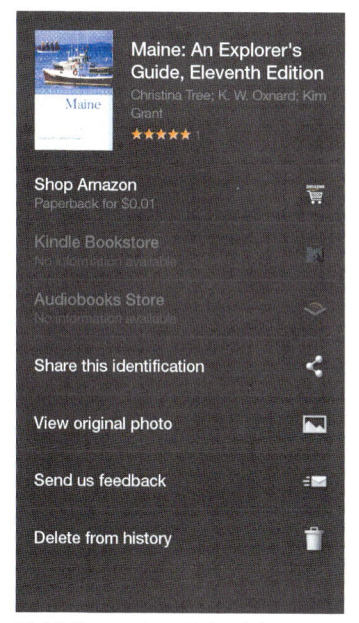

11.18 You can buy this book from Amazon, share it, or look at the photo Firefly took.

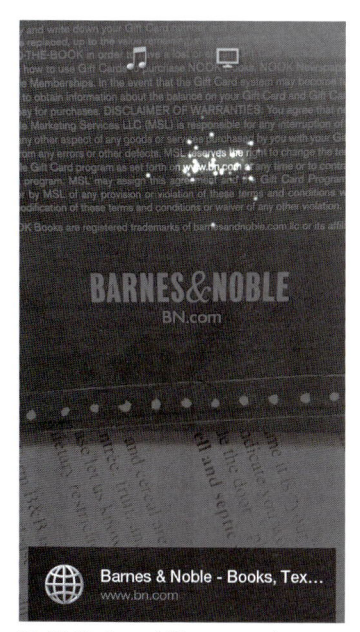

11.19 Finding a URL on a text-heavy document is easy for Firefly.

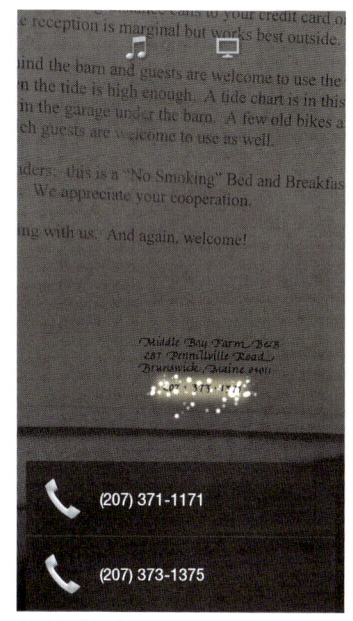

11.20 Telephone numbers can't hide either.

CHAPTER 11: CAMERA AND FIREFLY   167

Along the top of the Firefly interface are two icons: a music note and a monitor. Tap the note to identify a song that is currently playing in your location (11.21). The identification appears at the bottom of the scene and you can, once again, tap it to buy the track or see more information. The same thing applies for movies or TV shows. You don't actually have to point your camera at the screen; the movies and TV shows are identified based on audio, and information about the show is displayed (11.22). The neat thing about the TV and movie identifications is not only is the title identified but Firefly knows, for some shows and movies, which scene you're watching and shows you information about the actors currently on screen.

11.21 Firefly will tell you want that song is, and how much it costs on Amazon.

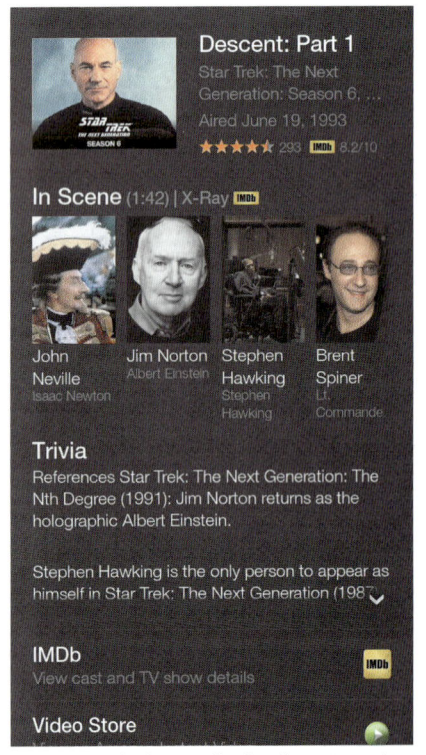

11.22 Not only does Firefly identify movies and TV shows, but it gives you some information about the actors in the scene you're currently watching.

168   THE AMAZON FIRE PHONE

The Firefly left panel lists all the Firefly-enabled apps on your phone (11.23). If the listed app has a checkmark next to it, like Flixter in 11.28, it is enabled for Firefly access. Tap it to uncheck it, and that app won't be able to use Firefly. At the bottom of the panel is a link to the Amazon Appstore so you can download more of them.

Tap the up arrow at the bottom of the Firefly screen to see your Firefly history (11.24). You can search your history or scroll through it. Tap an item to see the details. You can view the original photo that resulted in having this item in your Firefly history, or delete it all together.

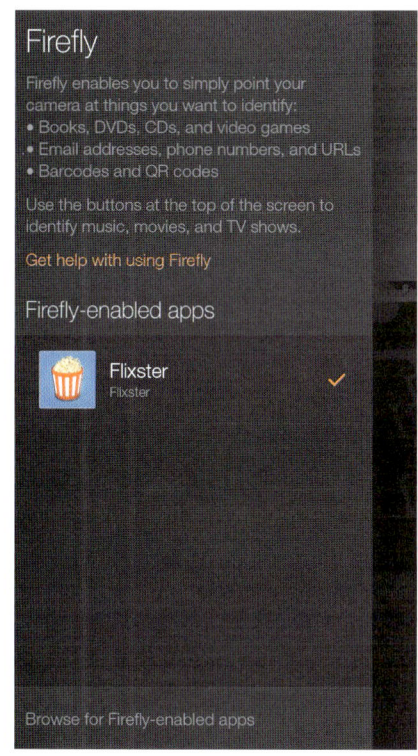

11.23 Firefly-enabled apps can be toggled on and off.

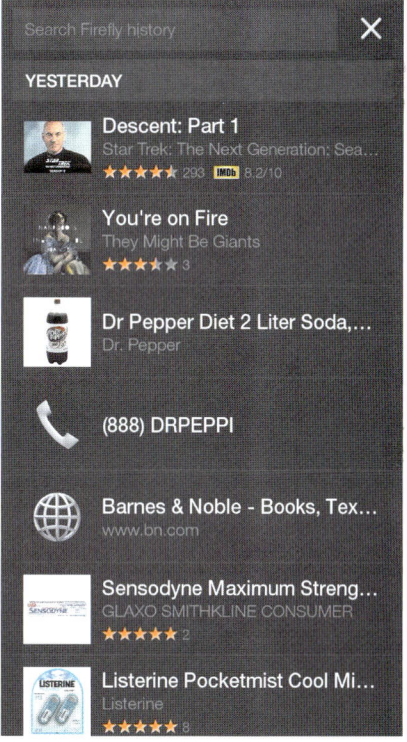

11.24 Your Firefly history is searchable.

Tapping Share this item will open the Sharing panel and share an Amazon link to the item, along with your name and "Shared from Firefly" (11.25).

11.25 Sharing links from Firefly includes a little advertisement for Firefly.

CHAPTER 12

Kindle

The Kindle app is where all your books live, and where you go to buy new ebooks from Amazon. This chapter covers how to read a book on your Fire, buying books, and how to add your own documents to your Fire for reading on the go.

Kindle Basics

Tap the Kindle icon, which is labeled Books in the home grid, to view your library of books (12.1). You actually have two Kindle libraries: Cloud and Device. The Cloud library includes books you've downloaded to your device (which have a checkmark on their cover) and those that are in Amazon's cloud waiting for download. By default, Kindle shows you the Cloud library, with the covers displayed in a grid. Swipe up or down to scroll through the list, or tilt the Fire back and forth to use auto-scroll.

Along the top of the display are two icons: search and the store.

Tap the search icon to search through all your books (12.2). Tap a book to download (if needed) and open it.

12.1 The Kindle app lists your books as a grid of covers.

12.2 Search your books with a tap.

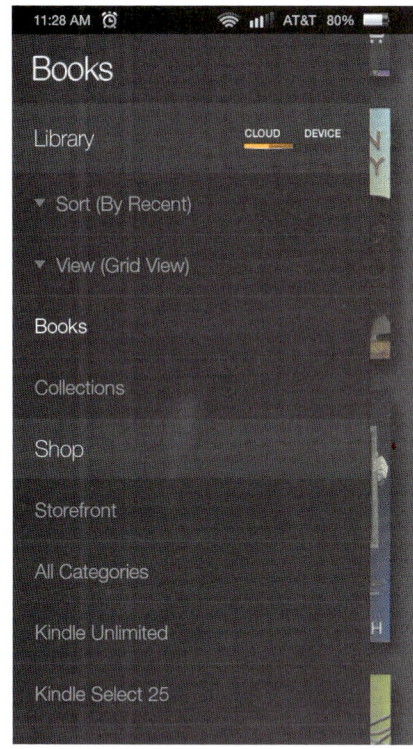

12.3 The left panel has some filtering options and allows you to jump to the store.

12.4 List view.

The left panel houses options to change your view of the Kindle library (12.3). List view includes more information about the book on the screen, but displays a much smaller cover (12.4). The current sort method is displayed and can be changed:

- By recent: Any book you've recently interacted with (downloaded, read, purchased) is displayed on top, with the rest ordered by the last time you interacted with them (or by purchase date if that's the only interaction you've had with that book).
- By title: Alphabetical by book title.
- By author: Alphabetical by author name.

To download a book to your Fire, tap it, and the download progress is displayed as an orange bar (12.5). Tap the X to cancel. As soon as the book downloads, you're taken to its first page.

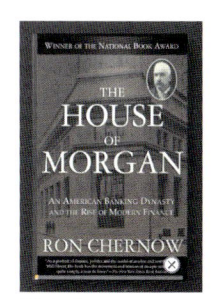

12.5 The orange bar tracks download progress.

CHAPTER 12: KINDLE   173

Long tapping a book reveals a few options: Pin to Home Grid, Download (if it is not on your device), Remove (if it is on your device), and Add to Collection.

Collections are ways of putting books together in groups. To create a collection:

1. Long tap a book you'd like to add to your new collection.
2. Tap Add to Collection.
3. A list of your existing collections appears, along with a New Collection button (12.6). If you want to add this book to existing collections, tap the checkmark next to all that apply. If you're creating a new collection, tap the New Collection button and give the collection a name (12.7).
4. Tap Add, and the book is added to the collection.

To view your collections, open the left panel and tap Collections (12.8). Long tap a collection to delete it or to pin it to the home grid. Tap to open the collection and see the books inside it. Tap the + in the upper-right corner to create a new collection right from here.

▶ **NOTE** Your Kindle collections are stored in the cloud, so they will sync with any hardware Kindles or other Kindle apps you might use.

12.6 Collections.

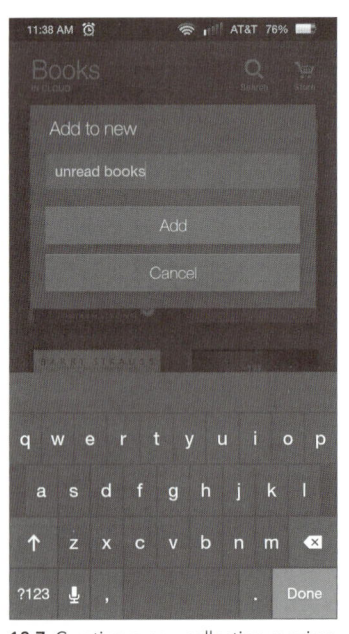

12.7 Creating a new collection requires you to name it.

12.8 The Collection view.

Reading

Tap a book to open it. You'll see the text of the book on your display, with some icons at the bottom and top of the screen (12.9). Tap or swipe to turn the page, and the menus disappear. Auto-scrolling isn't yet supported in the Kindle app, though Amazon is working on an update that will allow you to tilt your Fire to scroll through your books.

Tap the lower-left corner to cycle through the available progress options (12.10). The Fire can show you what page you're on, the location you're on, the estimated time it will take you to finish reading this chapter or the entire book, or nothing at all. Your progress will also be displayed on a banner on the book cover in the list view and on the Carousel (12.11).

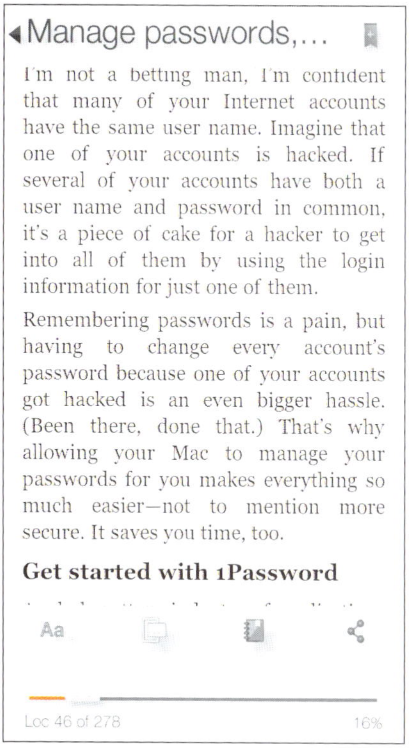

12.10 The various ways you can track your progress in a Kindle book. I usually use time left in chapter.

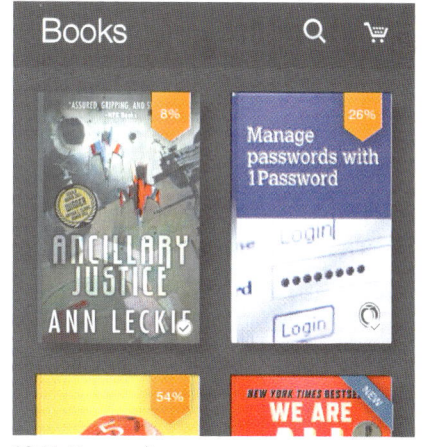

12.9 The Kindle reading screen.

12.11 Your reading progress appears as a badge on the book.

CHAPTER 12: KINDLE   175

Why is it so insecure? Well, even though I'm not a betting man, I'm confident that many of your Internet accounts have the same user name. Imagine that one of your accounts is hacked. If several of your accounts have both a user name and password in common, it's a piece of cake for a hacker to get into all of them by using the login information for just one of them.

Remembering passwords is a pain, but having to change every account's password because one of your accounts

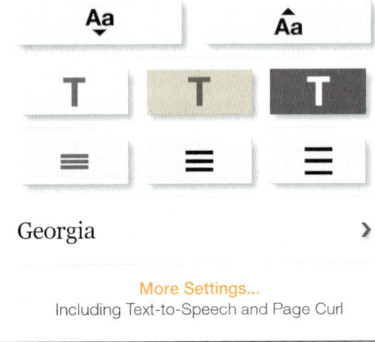

12.12 The appearance options.

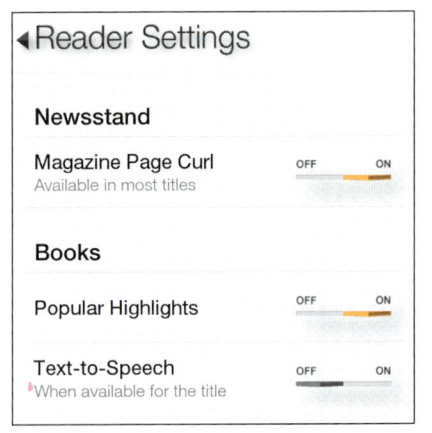

12.13 Reader Settings allow you to toggle off page curl and popular highlights.

Amaat willed it to be. Things happen the way they happen because the world is the way it is. Or, as a Radchaai would say, the universe is the shape of the gods. Amaat conceived of light, and

12.14 Popular highlights in action.

Tap the screen to bring the menus back. Tap the first icon to change the look of the text you're reading (**12.12**). You can change the font, the size, the margin, and the color of the background. As you tap options, the changes are immediately displayed onscreen.

Tap More Settings to turn on and off page-curl animation, popular highlights, and text-to-speech (**12.13**).

Popular highlights indicate passages in books that lots of Kindle readers highlight (more about highlighting in a bit) (**12.14**). The text is underlined, and the number of readers who have highlighted it is displayed. I find this fascinating, but you can toggle it off to de-clutter your screen.

When X-Ray is available for your book, the icon will be active. Tap this on a page to show all the characters, people, places, and terms mentioned on that page and where else in the book they appear (12.15). Tap an entry to see all the mentions of that character in the book, as well as a short blurb that gives you a sense of what the term, character, or place means in the book. Tap any of the mentions to go to that place in the book. The right panel also displays X-Ray information for books that have it (12.16). Tap the X or swipe left to close X-Ray and return to your book.

Tap Notes and Marks 📔 to see all the popular highlights, bookmarks, notes, and highlights in a book in one place (12.17). Tap any entry in the list to jump to the page with the note, bookmark, or highlight. Tap the X to return to the book's text.

Tap Share to bring up the Share panel so you can share a link to the book you're reading.

Under the four icons is a slider. Slide your finger along it to quickly go through the book both backward and forward.

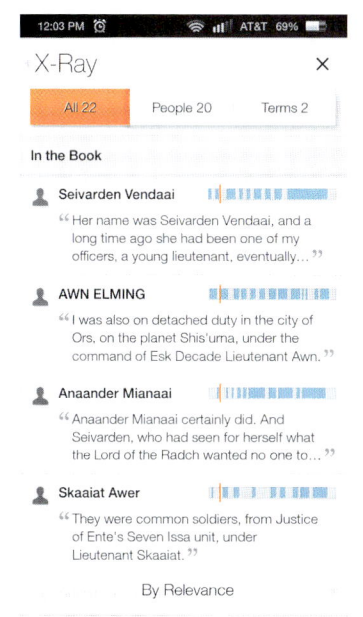

12.15 X-Ray lists all the characters, places, and important terms in a book in one place.

12.16 The right panel on X-Ray-compatible books shows information for the current page.

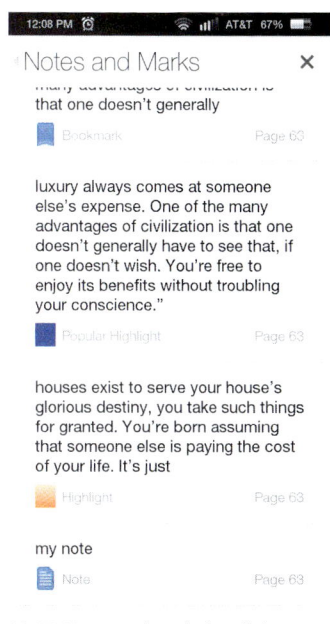

12.17 Notes and Marks list all the notes, bookmarks, and highlights in a book. Yours included.

CHAPTER 12: KINDLE   177

While you're reading a book, the left panel provides you with a way to jump from chapter to chapter or to a specific page (12.18). The book cover is displayed at the top. Tap the search icon to search within the book. Tap Go to Page or Location to jump to a specific point in the book (12.19).

▶ **NOTE** Some Kindle books have "real page numbers," meaning that Amazon has mapped out the page locations to match those of a print version of the book. Other Kindle books have Locations, which are basically precise line locations in the text.

Scroll down on the left panel to see the table of contents. Your current location is displayed in orange, along with the chapters in the book. Tap a chapter to go directly to it. Some chapter entries have an arrow next to them; tap it to see additional content you can jump to.

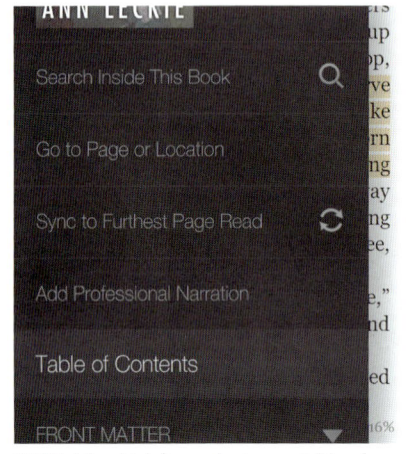

12.18 A book's left panel acts as a table of contents as well as a way to go to a certain page.

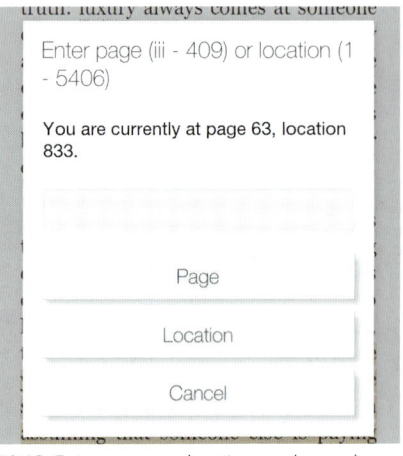

12.19 Enter a page or location number and tap the appropriate button.

Bookmarks, notes, highlights, and sharing

While you're reading, you might want to highlight a passage or make a note. To do this:

1. Long tap the text you want to note (12.20). A magnifier comes up.

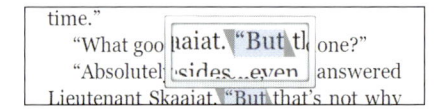

12.20 Selecting text in Kindle brings up the magnifier.

178   THE AMAZON FIRE PHONE

2. Move your finger to highlight the entire passage.

3. Lift your finger and a menu appears (12.21). Tap one of the colors to highlight the selected text. Tap the ✏ icon to bring up a text field for your note (12.22). Type your note, and tap Save.

The text is highlighted, and an icon appears on the passage (12.23). Tap the icon to read, edit, or delete your note.

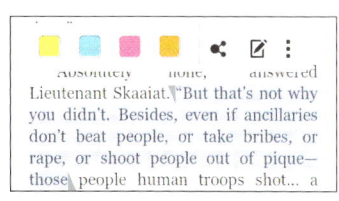

12.21 All the things you can do with selected text in the Kindle app.

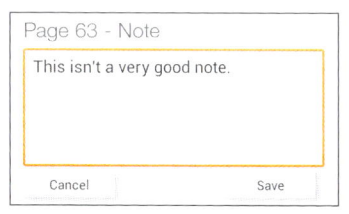

12.22 Notes can be as long, and as lame, as you want.

12.23 This text is highlighted and has a note associated with it.

To share a passage, do the same as above but tap the ⋖ icon. This brings up the Share panel, with the copied text and a link to the Kindle book passed along to whichever app you select.

There are two ways to bookmark a page in a Kindle book:

- Tap the upper-right corner of the screen while you're in the book. The corner turns blue, indicating that the page is now bookmarked (12.24).

- Tap anywhere on the display to bring up the menus. Tap the 🔖 icon and then tap Add Bookmark (12.25). A bookmark is added, and all your bookmarks are displayed.

12.24 The blue triangle indicates this page has been bookmarked.

To remove a bookmark, tap the 🔖 icon on a bookmarked page and then tap Remove Bookmark.

12.25 Tapping the bookmark icon lists all the bookmarks in the book, along with the option to create a new one.

CHAPTER 12: KINDLE 179

Lookups and translation

Come across a word you don't know in a Kindle book? Don't worry, just long tap the word and the definition will appear (12.26). Swipe up and down in the box to scroll through. That's the US English definition; tap the dropdown menu to change to another language. You can also tap Full Definition to read the entry in the dictionary. Swipe up to return to your book.

If you swipe left on the dictionary, you'll see the Wikipedia entry for the selected word (12.27). Swipe again and you can use Bing Translator to translate from one language to another (12.28).

12.26 Tap a word to look it up in Kindle's dictionary.

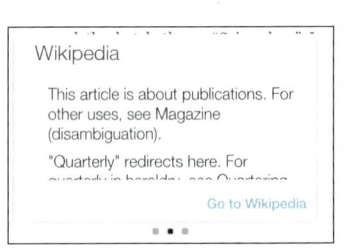

12.27 Swipe to see the word's Wikipedia entry.

12.28 Swipe again to translate it.

Kindle Store

Shopping in the Kindle Store is very much like shopping in the Appstore (12.29). Tap the icon to go to the Kindle Store. Books are highlighted in groups. The right panel displays a list of bestsellers, and the left panel allows you to jump to different sections of the store, which each display highlighted titles and groupings (12.30). Use Peek to see the price of the book and the average number of stars.

Tap a book to see its listing (12.31). Here you'll find the book cover (tap it or tilt your Fire to zoom in on the cover), the author's name, and the orange Buy button. The right panel lists other books bought by people who bought this book.

Swipe up to scroll down to the Download Sample button. Tap it to download a brief sample of the book, which will include a link to purchase the book. The book blurb, recommendations, and reviews (both editorial and from customers) come next, followed by the author's bio and the product details.

12.29 The Kindle Bookstore front page.

12.30 The left panel lists all the bookstore sections.

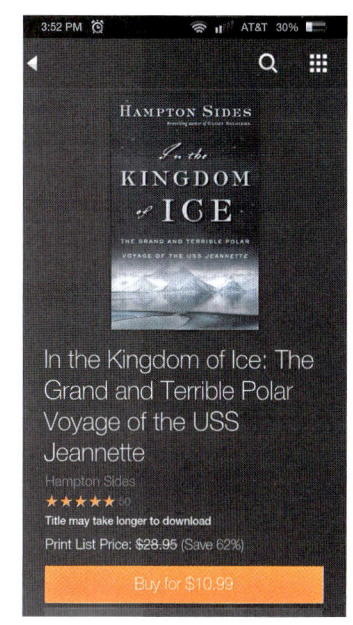

12.31 A book listing.

Tap the orange button to purchase the book. It starts downloading immediately, but if you change your mind, tap Cancel Order and you won't be charged (12.32). Once the download completes, the book appears in your library.

> **TIP** You can also buy Kindle books on Amazon.com and send them to your Fire. When buying a Kindle book, just select your Fire in the Send To list and it will show up when your Fire syncs.

12.32 The download progress is displayed after you purchase a book.

CHAPTER 12: KINDLE   181

Newsstand

Newsstand is an app dedicated to magazines and newspapers (12.33). Much like the Kindle app, it has a Cloud and Device library, grid and list views, and a store. Each magazine cover shows the number of issues you own. Tap to see the separate issues, and tap an issue to download and read it.

12.33 The Newsstand icon.

Magazines offer two ways to view content:

- Print replica makes the reader experience as close as possible to that of reading the actual print magazine (12.34). You can't change, select, or customize this view in any way.

- Text view gives you a more Kindle-like reading experience, with font controls, selectable text, and resizable type (12.35). Tap the screen and then tap ≡ to switch to this view. Tap the X to go back to the print replica view.

12.34 Print replica view includes the page curl.

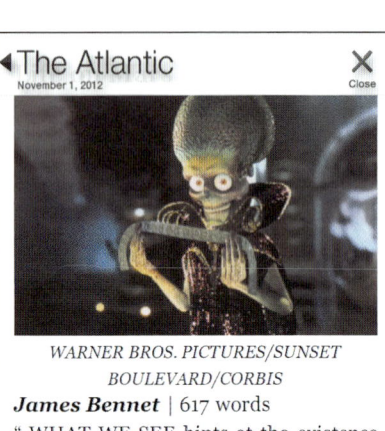

12.35 Text view is customizable.

▶ **TIP** The left panel shows the magazine's table of contents.

Tap the Subscribe Now button (12.36) to subscribe to a magazine (see Chapter 10 for information about managing your subscriptions). You can also purchase single issues.

12.36 Tap the orange button to subscribe to this magazine.

Docs

You can send personal documents to your Fire. They live in the Docs app (12.37).

Each Fire has an email address you can email documents to; they will automatically show up in the Docs app. To find out your Fire's address:

12.37 The Docs app contains your personal documents.

1. Launch the Docs app.

CHAPTER 12: KINDLE   183

2. Open the left panel, and tap Settings.

    The email address displays (12.38).

◄ Doc Settings

Send-to-Kindle Email Address
sm▓▓▓▓▓@kindle.com

12.38 Your Fire's personal document email address.

To get a personal doc onto your Fire, just attach the document to an email addressed to your Fire. The following attachment types are supported: Microsoft Word (.doc, .docx), HTML (.html, .htm), RTF (.rtf), text (.txt), JPEG (.jpeg, .jpg), Kindle Format (.mobi, .azw), and PDF.

After a few minutes, the document will appear in the Docs app ready for your reading pleasure. The Docs reader works just like the Kindle reader does.

CHAPTER 13

Music

Your Fire is a great way to listen to all your music on the go or at home. This chapter covers playing, buying, and organizing your music.

Playing Music

The first step to playing music on your Fire is to get some music onto it. If you've purchased any digital music from Amazon in the last couple of years, it will be available in your Amazon Music Cloud library (Amazon adds it for you automatically). If you don't have any music in your Cloud library, you can upload your music by going to the Amazon Music Cloud Player on Amazon's site and downloading the Music Importer.

Once you have music in your Cloud library, tap the Music app in the home grid to launch it (13.1). By default it shows you a list of albums in your Cloud library (13.2).

The left panel lets you display playlists, artists, albums, songs, or genres (13.3). Next to each heading is the number of items in it.

13.1 The Music app icon.

Tap an artist, album, or playlist to list all the songs in that group (13.4). Tap Play All to play all the songs on the list in the order in which they appear. Tap the ⤨ icon to play the list randomly. Tap individual songs to play them separately.

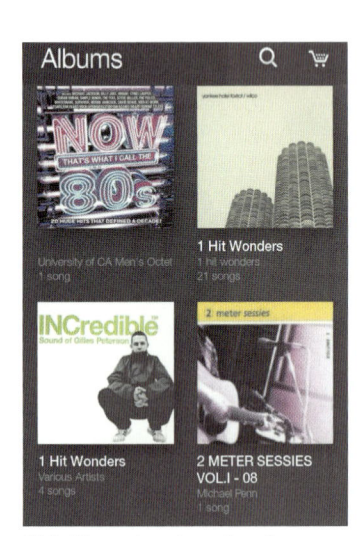

13.2 Album view shows the album artwork, the name of the album, the artist, and the number of songs.

13.3 The left panel allows you to navigate your Cloud and Device libraries as well as the Music store.

13.4 An album listing in the Cloud library.

As a song is playing, the play controls appear on the screen, along with the album art (if any) (13.5). Tap the pause button to stop the song, the left arrow to start from the beginning, and the right arrow to jump to the next song on the list (if there is one).

Tap the ↻ button once to play this song or list of songs over and over. Tap again to have it play twice and then stop ↻, and tap once more to turn off repeat.

Drag your finger on the slider to move through the song backward or forward.

▶ **NOTE** Playing songs from your Cloud library requires a network connection. You can download songs and albums by tapping the ⬇ button or by long tapping the song or album and tapping Download. Once the download is finished, those songs appear in your Device library and can be played without a network connection.

Some songs have a lyrics link above the play controls. Open the right panel as the song is playing, and the lyrics scroll by in time with the song, with the currently sung lyrics highlighted (13.6). Swipe up and down to scroll through the lyrics, and tap a lyric to jump to that point in the song.

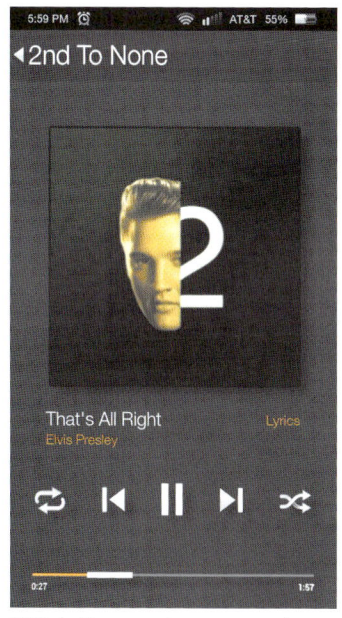

13.5 As the song plays, you see the familiar play controls.

13.6 The right panel displays, when available, the lyrics for the currently playing song.

CHAPTER 13: MUSIC   187

The song will continue to play if you move to a different section of the Music app. Controls appear at the bottom of the display so you can pause the music (13.7). If you leave the Music app, you can control the music in the notifications area of Quick Settings (13.8).

▶ **TIP** Press the button on the Fire's headphones to pause the currently playing song. Press again to start it.

13.7 As you go to different sections of the Music app, you can always see what's playing and pause it.

13.8 Controls appear in the notifications area as well.

Playlists

Playlists are the new mixtapes, except you can't give them to anyone. They allow you to group songs into that perfect mix that helps you do anything (like finish writing a chapter, for example).

Tap Playlists on the left panel to see what playlists you already have; Amazon Music syncs your playlists across devices (13.9). Amazon creates up to two playlists for you: Purchased, which includes all the songs and albums you've purchased from Amazon, and Recently Added to Cloud, which lists the music you most recently imported to the cloud (if you haven't imported any music, this playlist won't appear).

I have three on my Fire. Each has a name, the number of songs included, and an icon made up of four album covers from songs in the playlist. Long tapping the playlist I created brings up a few options (13.10).

Tap Add More Songs to bring up the music browser (13.11). Tap a song to add it, turning the plus into a minus. Tap it again and it will not be added. You can search for a specific song by typing a query in the search box, or you can add entire albums by swiping left to go to the album view. Tap the album you want to add, and then tap the album name to add it.

13.9 Playlists are mixtapes 2.0.

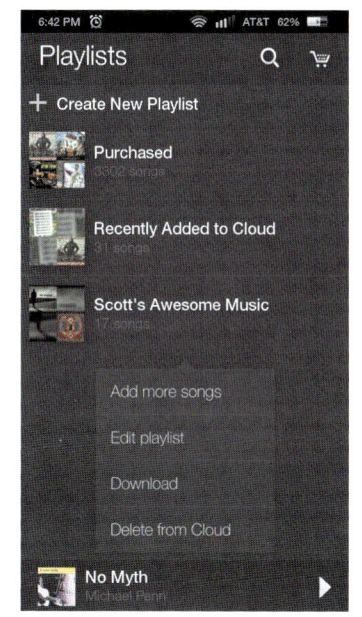

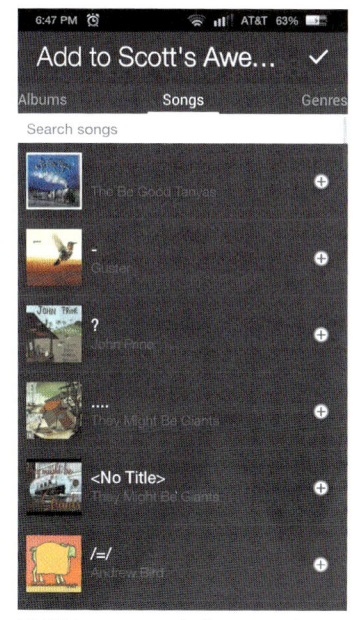

13.10 Long tap a playlist to see these options.

13.11 The music browser lets you quickly add music to playlists.

▶ **TIP** You can add albums and songs to playlists by long tapping them and tapping Add to Playlist. Select the playlist you want to add them to, and you're done.

Once you're finished adding things, tap the checkmark at the top right and the changes are saved.

Edit Playlist allows you to reorganize your playlist, change the name, remove songs from it, and add songs to it **(13.12)**. To move a song's position, press and hold on the three lines to the left of the song. This is the song's handle. While pressing down, slide up or down to drag the song to a new location.

Tap the minus sign next to the song to remove it, and tap the plus at the top right of the display to bring up the music browser so you can add more music to the playlist.

To change the name of your playlist, tap its name and edit away. If you tap the X in that field, all the text will be deleted and you can type a new name. Tap the checkmark to save your changes.

Tap the Download button to download all the songs on the playlist to your device for offline play. Tap Delete from Cloud (or Device) to delete the playlist.

To create a new playlist, tap the New Playlist button at the top of the playlist screen (13.13). Give the playlist a name and tap Create, and you're taken to the music browser to add music to your new playlist.

13.12 Arrange the order of songs and remove unwanted songs from your playlist here.

13.13 A good name is important for a playlist.

Managing Music

When you long tap a song or album, some options appear.

- Explore Artist: Want to see more from this artist? Tap this and you'll be taken to the artist's page in the Music store.

- Go to Album: When you're looking at a song, tapping this takes you to the full album.

- Delete from Cloud/Device: Deletes the song, or album, from either your Fire or the cloud.

Purchasing Music

The Music Store displays highlighted and recommended music on the front page (13.14). Tap one to see either the list of music included in the promotion or more information about that album or song.

The left panel has all the sections of the store and lets you get back to your own library quickly (13.15).

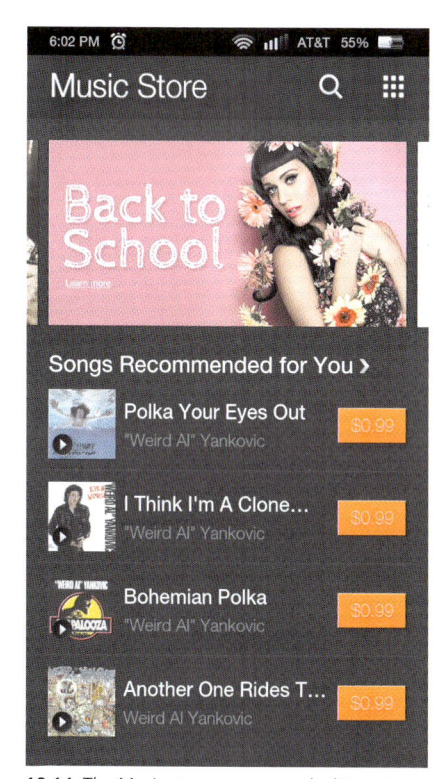

13.14 The Music store recommends things it thinks you might like. Weird Al rocks.

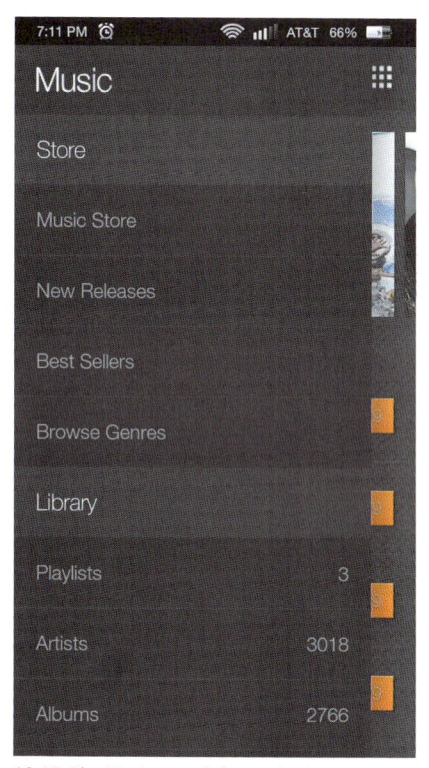

13.15 The Music store left panel is how you get from section to section.

Once you've found a song or album you're interested, in the listing will give you some information about it (13.16). The album artwork is displayed, along with the title, the artist name, the average star rating, and a Buy button. The songs are listed below in a list. Scroll down to see product details (release date and the like), reviews, related albums, and a button that links to more music by the same artist.

You can preview songs by tapping the play icon next to a song (or on the album to play all the previews in order). You'll hear a 30-second clip to judge the song by.

You can tap the Buy Album button to purchase the entire album, or the button with a price next to the song you're interested in. Some albums don't allow you to purchase songs separately.

Tap the Buy button twice to confirm your purchase. The Play or Download button appears (13.17). Tap it to set your automatic download preference (13.18). By default, purchased music is added to your Cloud library but not downloaded to your device. Tap Play to stream the music to your Fire, or Download to save it on the Fire.

13.16 An album on the Music store.

13.17 Tap the Play or Download button to listen to your newly purchased music.

13.18 You can set your download preferences right here.

CHAPTER 14

Instant Video

Amazon has an extensive selection of TV shows and movies that you can rent or buy, all of which will play on your Fire (14.1). You can also load your Fire up with videos of your own. This chapter covers using the Instant Video app to rent and buy videos, watching videos on your phone and Fire TV, and getting your own videos onto the device.

14.1 The Instant Video icon.

Renting/Purchasing/Downloading

Most of the video you watch on your Fire uses the Instant Video app. Tap the app to launch it, and you'll see a selection of movies and TV shows that are available in the Video Store (**14.2**). Many of the featured videos will have a banner at the top that says Prime. Streaming these videos to your Fire is free if you're an Amazon Prime member (more on Amazon Prime in Chapter 16); if you aren't a Prime member, you'll need to rent or buy them.

Swipe up and down to see all the featured shows and movies. Swipe left and right in the groupings to view all the entries.

The left panel includes links to different areas in the store, including Top TV Shows and Top Movies (**14.3**). The right panel lists movie and show recommendations based on your viewing habits (I like *Star Trek*) (**14.4**).

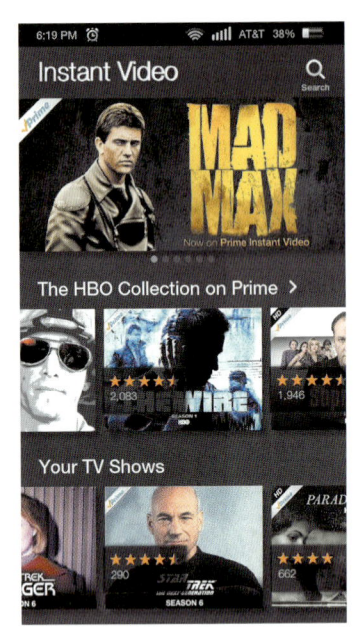

14.2 The homepage highlights videos you might like.

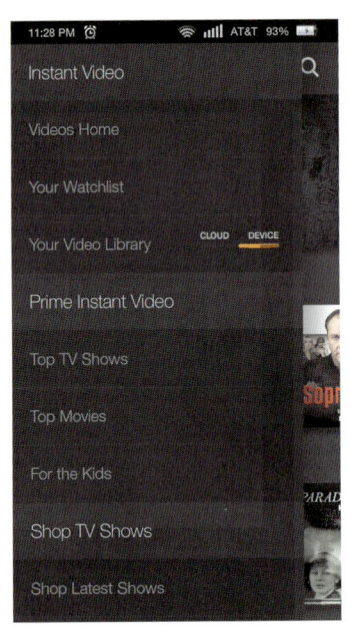

14.3 The left panel has navigation options.

14.4 The right panel lists movies and TV show recommendations.

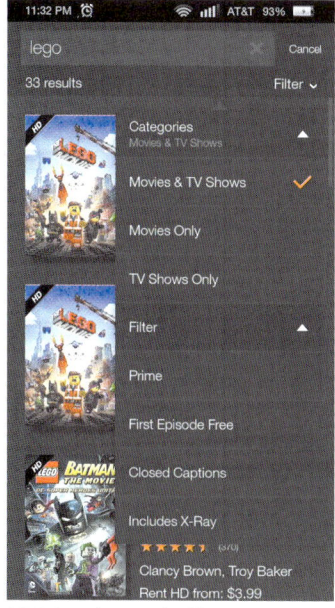

14.5 Searches can be filtered.    **14.6** A movie listing. 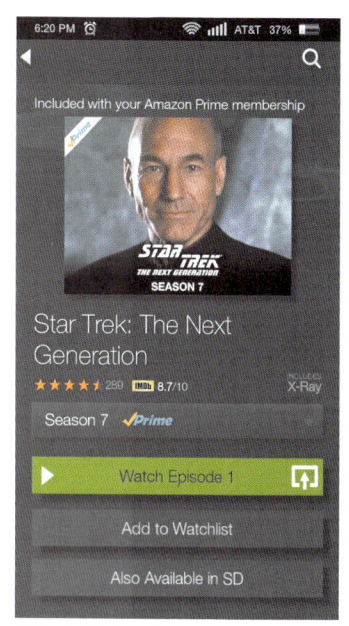   **14.7** A TV show listing includes seasons.

If you can't find what you want on the home screen, in the right panel, or in the sections listed on the left panel, search the store by tapping the 🔍 icon and entering a query. Tap a suggestion in the list, or tap the search key to see the search results. Filter the results by tapping Filter and selecting filtering criteria (14.5). For example, if you want to see only TV shows that are available via Prime and include X-Ray (more on this in a bit), tap those criteria in the filter list and the results will show matching items.

Once you've found what you want to watch, tap it to go to its details (14.6). The poster art is featured prominently, along with Rent, Buy, and Watch buttons, depending on the video you're looking at.

TV show listings are slightly different from movies since shows have more than one season (14.7). Tap the Season menu to switch seasons. Each season lists the episodes that are available to watch.

When you rent a video from Amazon, you have 30 days to watch it. Once you start watching it, however, you must finish watching in a set amount of time (usually 24 hours) or you'll have to rent it again. You can watch it as many times as you like during that time. If you buy a video from Amazon, you can download it to, or stream it on, any compatible device (which includes your Fire) forever. It is yours to watch as many times as you like, over and over again.

To rent or buy a video, tap the appropriate button. Tap again and payment is processed. A Watch Now button appears on the video (14.8). Rentals also include a note letting you know how long your rental period is and when it expires.

14.8 Tap Watch Now to watch a video.

Tapping Watch Now on any video will stream that video to your Fire phone. You can also download rentals and purchases by tapping the Download button. This takes longer initially, since the video files are large, but you can watch downloaded videos without a network connection, whereas streaming requires a constant connection.

Prime shows and movies are usually streamed, but you can download them to your phone by tapping the ⬇ icon.

Watching

You can watch a video by finding it in the Video Store and tapping the Rent, Buy, or Watch button. But what about the videos you've already rented or bought? Those are in your Video library (14.9). To get to it, open the left panel and tap Your Video Library.

Tap a video and you're taken to its listing. Tap the Watch Now button to play it. No matter what kind of video you're watching—rented, purchased, or Prime—the video controls are the same (14.10). The controls are visible when the video starts to play, and they fade after a bit. Tap the screen to bring them back. From left to right, the controls on the bottom of the screen are:

- Pause/play button: Tap this while your video is playing to pause it, and tap it when the video is paused to resume playback.

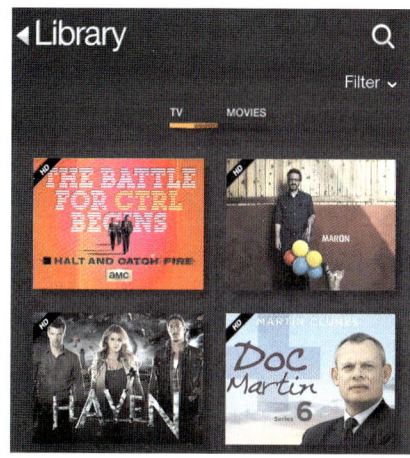

14.9 Videos purchased from Amazon are in your Cloud library.

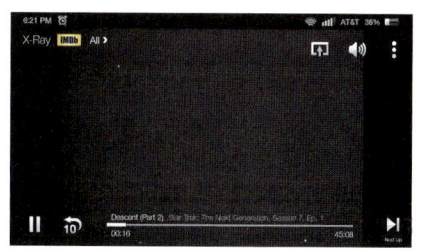

14.10 The video play controls.

- Skip back: Tap this button to skip back 10 seconds in the video. Tap it repeatedly to hop back in 10-second increments.

- The timeline: Atop the line is the name of the video you're watching. Below that is the timeline of the video, with the playhead at the current point in the video. Drag the playhead to another place on the timeline to jump the video to it. To the left of the line is the timecode of your current position in the video, and to the right is the total amount of time in the video.

- Next Up: When you're watching a TV show, tap the Next Up button to skip to the next episode in the season. Great for binge-watching.

The top controls, from left to right are:

- Play on: Streams video from your Fire to supported devices. This is covered in the next section of this chapter.

- Volume: Tap this to turn up or lower the volume.

- Menu: When tapped, this expands to show:

    - Captions: Some videos support closed captioning. Tap this to both enable captions and customize the look of the captions (14.11). Turn on captions to preview their look.

    - Zoom: Tap to view a video full screen.

    - Exit Playback: Done with your video? Tap Exit Playback and you're back at the video's details.

In 14.10, you can see the X-Ray button in the upper-left corner. This appears on X-Ray-enabled videos. Tap it, and a menu appears with information about the scene and the actors in it (14.12). Tap an actor to see more information about them.

14.11 Customizing captions.

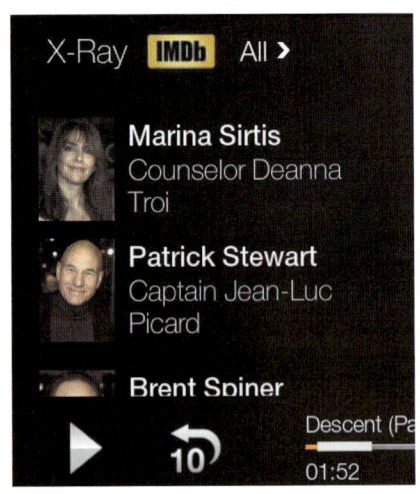

14.12 X-Ray details from IMDb.

Second screen

If you have the proper equipment, it is possible to play videos from your Fire on your TV and use your Fire as a second screen. Amazon calls this *flinging* video. To do this, you'll need an Amazon Fire TV, PlayStation 3 or 4, or a recent Samsung TV.

1. Ensure that your Fire is on the same Wi-Fi network as the device you want to fling your video to.

2. When you play a video, you'll see the icon. Tap it and a menu will appear listing the devices that can receive the video (14.13).

3. Tap the device you want to play the video on. The video will play, and your Fire will become a secondary screen displaying X-Ray information, if available, along with play controls (14.14).

   To stop playing the video on the other device and resume on your Fire, tap again and then tap your Fire's name.

14.13 Select a supported device to play the video.

14.14 Your Fire acting as a second screen.

Watchlist

As you're browsing the Video Store, you'll come across movies and shows that you want to watch eventually but not now. Long tap them and tap Add to Watchlist, or tap the Add to Watchlist button in the entry listing. Your watchlist is a list of videos you want to see. Access it by opening the left panel and tapping Your Watchlist. Tap any entry to watch it. Long tap it and tap Remove from Watchlist to delete it from the list.

Your Own Videos

To transfer your own videos to your Fire, you'll need to connect it to a computer using a mini-USB cable. See Chapter 10 for details about this process. The video format must be in the Fire's supported list: MPEG4, VP8, H.264/MPEG4/AVC, MPEG4 SP, H.263, AVI, and HDCP2.x. Simply move the video file into the Fire's Movies directory. Wait a few seconds, and go to your Device's Video library (Instant Video > left panel > Personal Videos).

Video Settings

Open the left panel, swipe up, and tap Settings to customize the Instant Video app (14.15). Most of the settings have to do with how the app works when you're on a cellular connection:

- Mobile Network for Video: On by default, this allows you to stream shows and movies while you're on your Fire's data plan. Most of these plans have a limit, and video can chew through data. Tap this setting and change it to Off to stream video only while you're on Wi-Fi.

- HD Download Quality: When you download a video you'll be asked what version you want: SD (Good), 720p (Better), or 1080p (Best). The better the quality, the bigger the video file. If you want to choose a default, tap this setting and pick from the list.
- SD Download Quality: The same as above, but with videos in standard definition.
- Dolby Digital Plus 5.1: You can turn this on (the default) or off. With it on, your videos sound better but use a little more bandwidth and storage, because more sound data is included.
- Clear Video Search History: Does just what it says.
- Signed in as: This is the account you're signed into on Amazon Instant Video. By default, this is the same account that your Fire is linked to, but you can use a different account for video than for the rest of your Fire. Tap here and enter the credentials for a different Amazon account to use that account for videos.

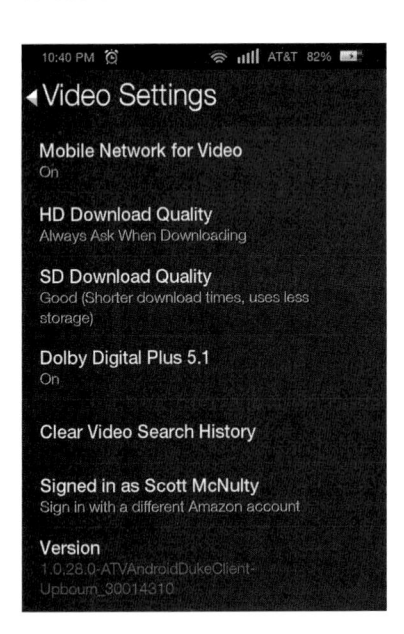

14.15 Video settings.

CHAPTER 15

Voice Control

You can perform some tasks on your Fire using just your voice (and a finger to press the home button). Amazon is the first to admit that the Fire's voice controls aren't comparable with those on competing smartphones (Apple's Siri, Google's Google Now, and Microsoft's Cortana), but they are working on adding functionality. As of this writing, you can use your voice to dictate (covered in Chapter 3), call someone, send texts and emails, and search the web or Amazon.com. This chapter covers all of these uses.

▶ **NOTE** Voice control requires a network connection.

Make Your Fire Listen

Before you talk to your Fire with the expectation that it'll do anything, you need to make it listen. Press and hold the home button until the screen says Hello and some fireflies around a microphone icon (15.1).

Now your Fire is listening to you. If you're in a loud room, your Fire might not be able to figure out what you've said. It will display an error message (15.2). Tap the microphone icon to try again.

15.1 Voice control.

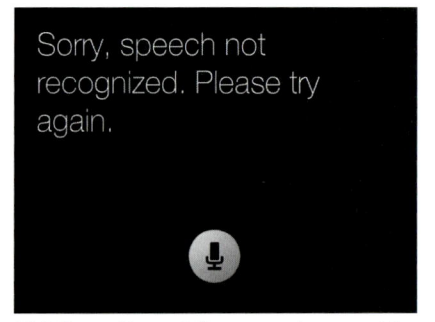

15.2 If your Fire can't figure out what you said, this message appears.

Phone Calls

To make a phone call with your voice:

1. Activate voice control.
2. Say "Call [name of contact]."
3. If that contact has more than one number, your Fire will ask you which number you would like to call (15.3). Say "home" or "mobile."

   Your Fire tells you that it is going to call your contact, and the call is placed (15.4).

202   THE AMAZON FIRE PHONE

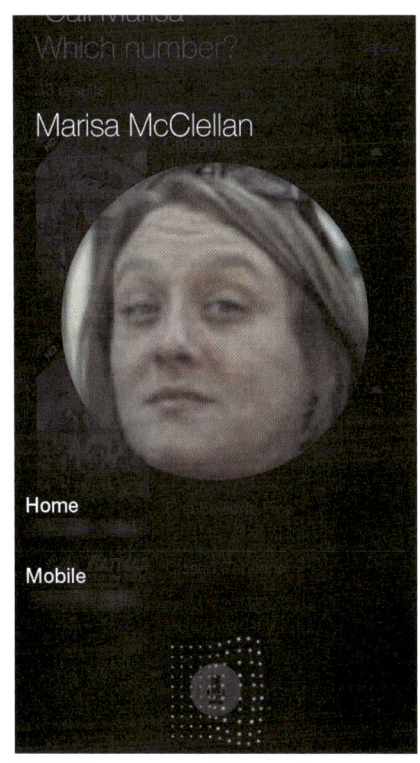

15.3 Calling a contact with more than one number.

15.4 Placing a call.

Messaging

To text or email someone with your voice:

1. Activate voice control.
2. Say "text [name of contact]" or "email [name of contact]."
3. If you have more than one contact with the same name, your Fire will list them all. You can tap the correct one, or you can tell your Fire which one on the list to message (say, "the first one" or whichever one you're after).

CHAPTER 15: VOICE CONTROL   203

Once the contact is identified, you'll have to pick which email address you want to send to (texts are automatically sent to mobile numbers). Once again, you'll see a list of choices. You can just say "the second one" and the Fire will know what you mean.

4. For emails, your Fire will ask what you want the subject line to be. Then dictate the content of your message (text or email). You'll see a preview (15.5). Say "edit" if you want to change something, or say "cancel" to discard the message.

15.5 Previewing a message.

> **Send it?**
>
> To: Marisa McClellan
>
> Hello
>
> I sent you this email from my fire.

5. Once you're happy with your message, say "send" and off it goes.

▶ **TIP** You can also activate voice control and say, "Text Marisa I'll see you soon." You'll see a preview of the message. Say "send" to send it.

Searching

1. Activate voice control.
2. Say "search the web for [something]" or "search Amazon for [something]."

   Your Fire will say, "Let me get that for you" and after a moment will take you to either Silk search results or Amazon search results.

   You can also just say "search the web" or "search Amazon." Your Fire will ask you to say what you're looking for and then perform the search.

CHAPTER 16

Amazon Prime

Amazon Prime is a membership program that includes shipping benefits, Prime Instant Video, Prime Music, the Kindle Lending Library, and Kindle First.

Prime costs $99 a year, but your Fire (for a limited time) comes with a free year of Prime membership. If you're already a Prime member when you buy a Fire, your membership is extended by a year. This chapter gives you an overview of Prime's benefits and how to access them with your Fire.

Shipping

Amazon Prime started off by offering free two-day shipping and $2.99 overnight shipping for select physical items on Amazon.com (denoted with a Prime badge) (16.1). People loved it, and Amazon has expanded the shipping benefit to include $5.99 same-day delivery in some cities and a host of additional benefits.

16.1 The Prime logo means this book is eligible for free two-day shipping.

Share the shipping

You can share your Amazon Prime shipping benefits with up to four people. Amazon says they must be household members, but they don't verify it.

To share Prime shipping:

1. Go to www.Amazon.com/prime, click Manage Prime Membership, and log in to your Amazon account.

2. Scroll down until you see the Invite a Household Member section at the bottom of the page. Click to expand (16.2).

16.2 Adding household members to your Prime account gives them free shipping.

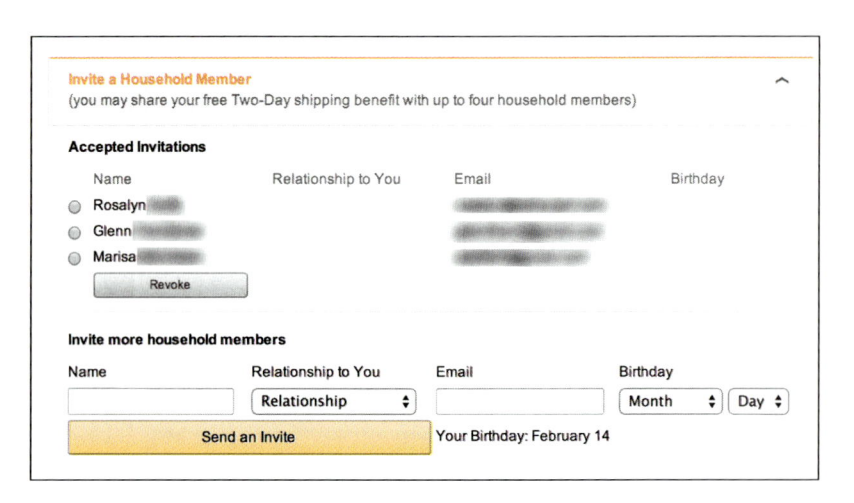

3. Enter the person's name, relation to you, email address, and date of birth. Click Send an Invite and they are invited to join your Prime shipping family.

   Keep in mind that this shares only the shipping benefit; this person will not be able to use any of your other Prime benefits.

Prime Instant Video

Watching video on your Fire is covered in Chapter 14. As you browse the video selection in the Instant Video app, you'll see a number of shows and movies that display a Prime banner (16.3). These are all part of the Prime Instant Video benefit.

Think of Prime Instant Video like Netflix: you can stream (or download to your Fire) any movie or show in the program without paying additional fees. Prime videos are also viewable on Amazon.com, on a Fire TV, and on a host of other devices.

16.3 Prime Instant Videos have the Prime banner in the upper-left corner.

Prime Music

Prime Music is the latest addition to the Prime program. It is a music-streaming service akin to Pandora or Spotify. Amazon has many songs and albums that you can listen to, in their entirety, for free if you are a Prime member.

The Fire app that lets you take advantage of Prime Music is already in your Cloud library (16.4). Tap it to download it to your Device library, and tap it again to launch it.

16.4 Amazon Prime Music.

You'll be logged in with the Amazon account associated with your Fire. Tap Explore Amazon Prime Music to jump into the available music (16.5). The default view is of the Amazon Prime Music song list; tap the Albums or Playlists tab to switch to that view. Think of Amazon Prime Music as the Music Store with one exception: you don't have to pay for the music if you're a Prime member.

All the items in the list have a 30-second preview. Tap the play button to hear it. Tap an album to see all the songs on the album (16.6). You can also purchase a song or album here if you want to make sure it is yours forever (songs can be removed from Prime Music by Amazon).

When you find a song, playlist, or album you want to hear in its entirety, you have to add it to your Music library. Tap the Add button.

Your Cloud and Device libraries are available in the Prime Music app via the left panel (16.7). It works just like the Music app in Chapter 13; the only difference is the addition of the Prime Music and Prime Playlists categories.

▶ **NOTE** Prime is a streaming service, so it requires a network connection.

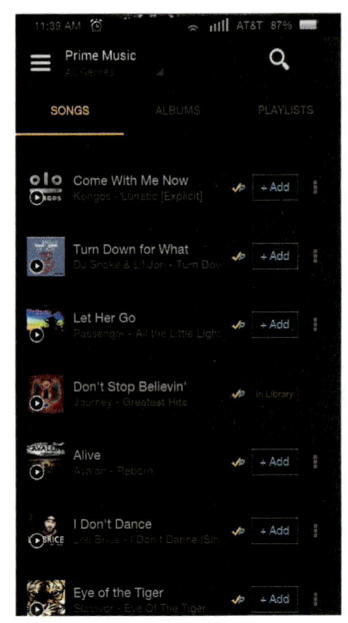

16.5 Prime Music has lots of music you can add to your library for no additional cost.

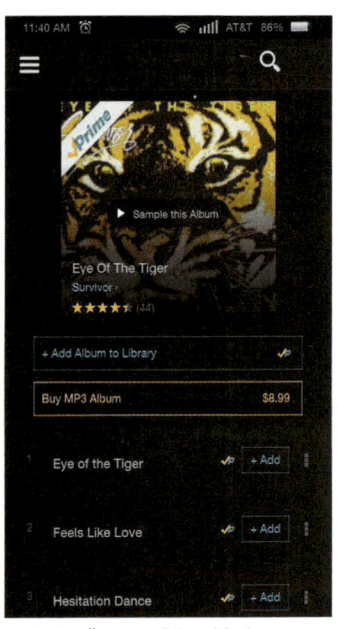

16.6 An album on Prime Music.

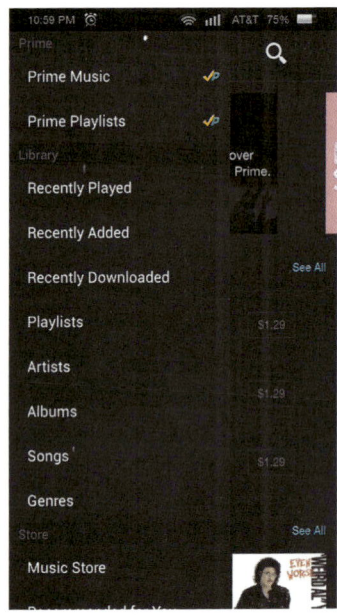

16.7 Prime Music and Prime Playlists.

208 THE AMAZON FIRE PHONE

Kindle Lending Library

Prime members can borrow one book a month from the Kindle Lending Library. You can keep that book out for as long as you like, but you can only have one book out at a time (and take out one book a month).

The Kindle app's left panel includes a listing for the Kindle Lending Library. Tap it to see the list of eligible books (16.8). Tap a book that looks interesting, and you'll see the Borrow for Free button, as well as the Prime logo (16.9).

As you are browsing the Kindle Store, the Prime logo lets you know that a book is included in the Kindle Lending Library.

Tap Borrow for Free to download the book. If you already have a book out, you'll need to return it first (16.10).

▶ **NOTE** Amazon has introduced a new service, called Kindle Unlimited, that lets you borrow as many books as you like from over 600,000 titles. It costs $9.99 a month, in addition to your Prime membership. Read more about it at www.amazon.com/KindleUnlimited.

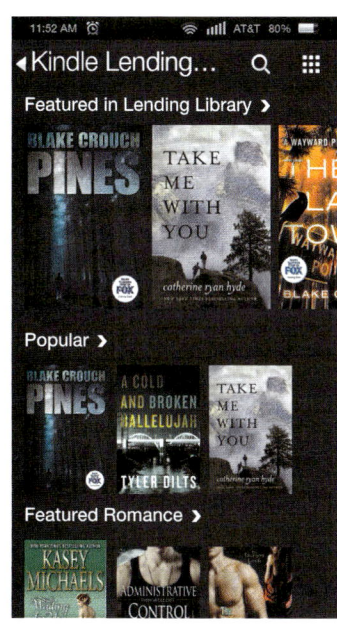

16.8 You can borrow one book a month from the Kindle Lending Library.

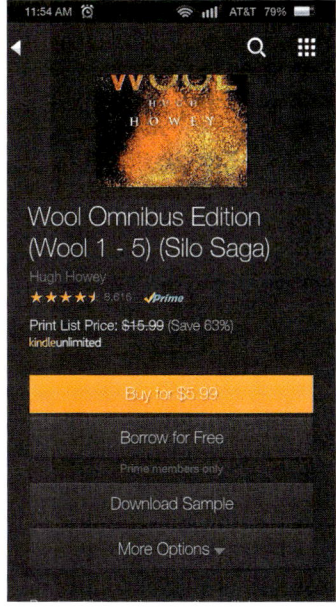

16.9 Tap Borrow for Free to borrow this book.

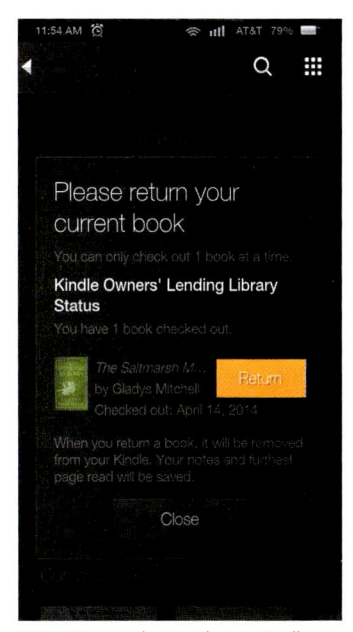

16.10 You can have only one Kindle Lending Library book out at a time.

CHAPTER 16: AMAZON PRIME   209

Kindle First

Kindle First is another book-centric benefit from Amazon Prime. Every month, Amazon editors choose four upcoming books for the Kindle First list. Prime members can download one of those books for free and read it before its official publication date.

The left panel in the Kindle app has a Kindle First entry. Tap it and you're taken to this month's selections (16.11). Tap them to read about them, and then tap the Buy Now for Free button to get your copy (16.12).

16.11 Kindle First gives Prime members a sneak peek at new books.

16.12 Download a free Kindle First book.

CHAPTER 17

Security

The more personal information we store on our smartphones, the more useful they become. That's why securing your Fire is so important. If your phone is misplaced or stolen, would you want someone poking around in your apps, contacts, messages, and browsing history, let alone buying things using your Amazon account?

I certainly wouldn't. This chapter shows you how to enable a passcode and encryption on your Fire and how to manage your Fire remotely using tools on Amazon.com.

On Your Fire

The first thing I do when I get a new phone is set a PIN (personal identification number) for my lock screen. So when someone tries to unlock my phone, they'll need to enter the PIN first.

The Fire supports either a PIN (four digits) or a password (numbers and letters). Here's how to enable it:

1. Open Quick Settings by swiping down, and tap Settings.
2. Tap Lock Screen, and then tap Set a Password or PIN.
3. Select a lock method from the list (17.1). The Fire defaults to None. PIN sets a 4-number code, and Password lets you set a password with numbers, letters, and symbols (it has to be at least four characters).
4. Enter the PIN or password twice to set it.

17.1 You can choose a PIN or password to lock your phone.

Now when you unlock your screen, you'll be prompted to enter your password or PIN (17.2).

When you have a PIN or password set, you'll need to enter it before you can change the Lock Screen options or the PIN or password itself.

▶ **NOTE** You can also set your lock screen background by going to Settings > Lock Screen > Select a Lock screen Scene.

Your Fire will lock itself after 10 minutes of inactivity. To change this amount of time, go to Quick Settings > Settings > Lock Screen > Change the Automatic Lock Time. You can set the time to 30 seconds or 1, 5, 10, 30, or 60 minutes (17.3). By default, pushing the power button to put your phone to sleep also locks it; toggle this off if you don't like this behavior.

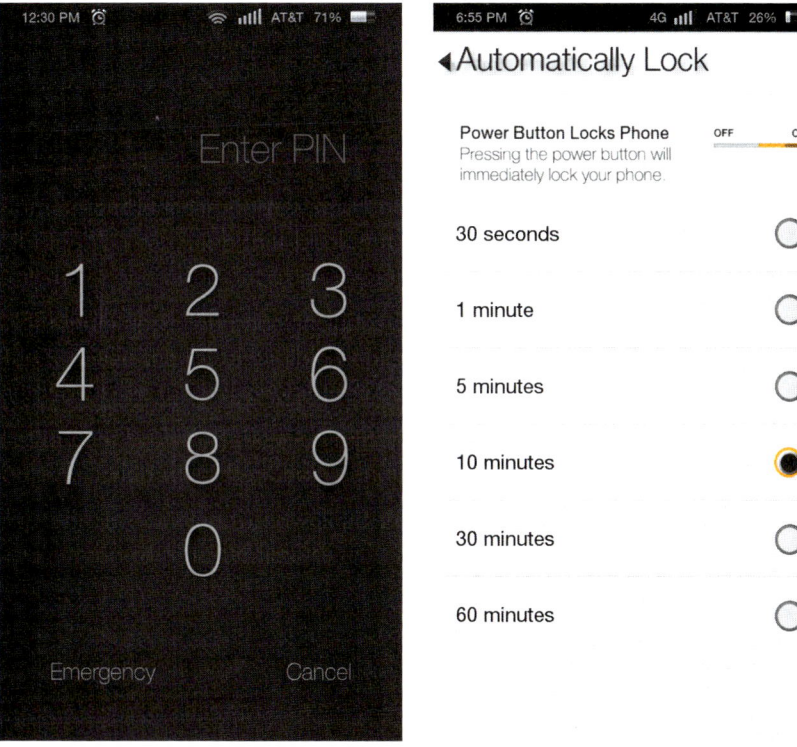

17.2 A locked Fire requires a PIN (or password).

17.3 The shorter the delay, the more secure your phone.

Encryption

A lock screen password is a good first defense, and encryption is a great next step. Encryption makes it impossible for others to read your data unless they have the encryption password you set. This means that if you lose your phone and a hacker hooks it up to a computer in an attempt to grab your data (since you have a lock screen password, they can't just get in), they will be thwarted.

Once you turn on encryption, you'll have to enter a password whenever your Fire starts up. This is in addition to your lock screen password, but it is only needed when the Fire starts up.

To encrypt your phone:

1. Open Quick Settings, and tap Settings.
2. Tap Device, and then tap Manage Enterprise Security Features.
3. Tap Encryption, and you'll see some text explaining it (17.4).
4. Plug your Fire into its charger (since the process can take an hour), and tap Encrypt.
5. Set a password, and tap Continue (17.5). The Fire will reboot, encrypt itself, and start up again.

   Now when you start your Fire up, you'll need to enter your encryption password (17.6).

17.4 Your Fire has to be plugged in before you can encrypt it.

17.5 Tap Continue to encrypt your phone.

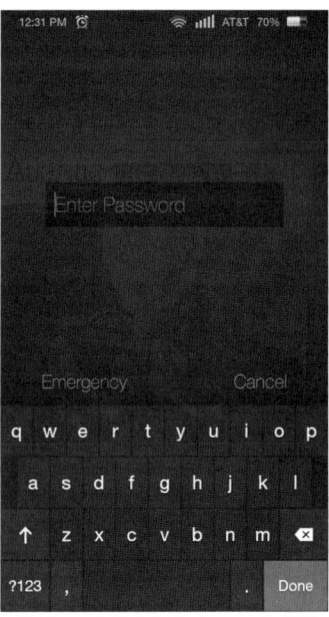

17.6 An encrypted phone requires a password to start up.

Updates

One of the best ways to make sure your Fire is secure is to keep its software up to date. Apps are automatically updated (see Chapter 10), but your Fire needs some manual intervention when a *system* update is available.

You'll get an alert telling you an update is ready. Tap it, and you're taken to the System Updates screen (you can also get there by going to Quick Settings > Settings > Device > Install System Updates) (17.7).

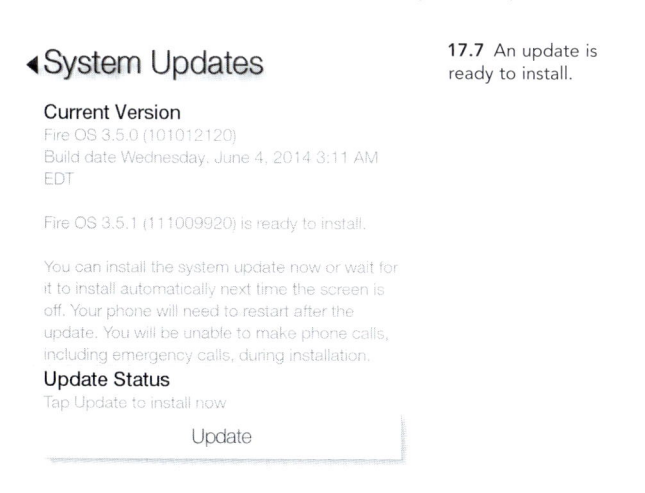

17.7 An update is ready to install.

The version of Fire OS on your phone is displayed, along with the latest version. If the version on your Fire is lower than the available version, there will be an Update button. Before you tap it, make sure your phone is plugged in. Tap Update and the update downloads, reboots your phone, and updates the system.

> ▶ **TIP** You can manually check for updates by going to Install System Updates and tapping the Check Now button.

Amazon.com

You can manage a few things about your Fire on Amazon.com. Go to www.amazon.com/manageyourkindle, and log in with your Amazon account.

Click Your Devices, and select Devices from the Show menu (17.8). Your Fire will be in this list. Click it, and you'll see some information about your Fire, including its email address and serial number.

17.8 Managing your Fire on Amazon.com.

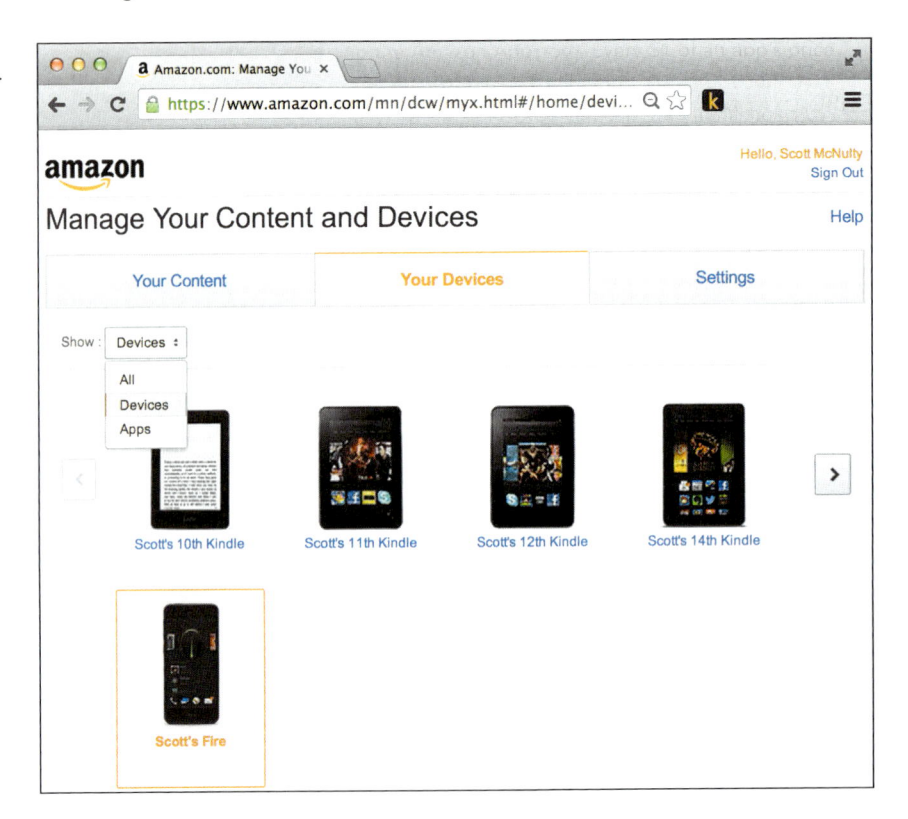

Deregister

Your Fire is registered to your Amazon account. If your phone is ever lost or stolen, or you sell it, you can deregister it by clicking the Device Actions menu and choosing Deregister (17.9). A warning pops up letting you know what deregistering means (17.10). Click Deregister, and no one will be able to purchase things on your Fire until you re-register it.

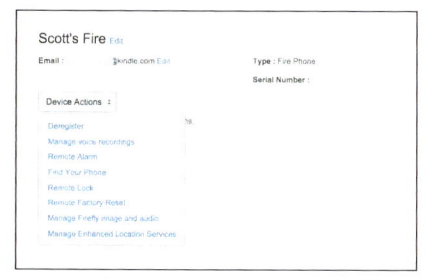

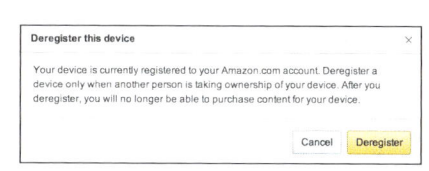

17.9 Fire details.

17.10 Amazon.com allows you to remotely deregister your Fire.

Locate on map

If your your Fire is lost or stolen, you can locate it on a map, provided that the phone is turned on and has a network connection. Choose Find Your Phone from the Device Actions menu.

You get a warning that the process might take up to 30 seconds. After a few moments, a map with a circle depicting your Fire's location is displayed (17.11). Keep in mind that the location isn't pinpointed exactly; this just gives you a general idea of where your phone is.

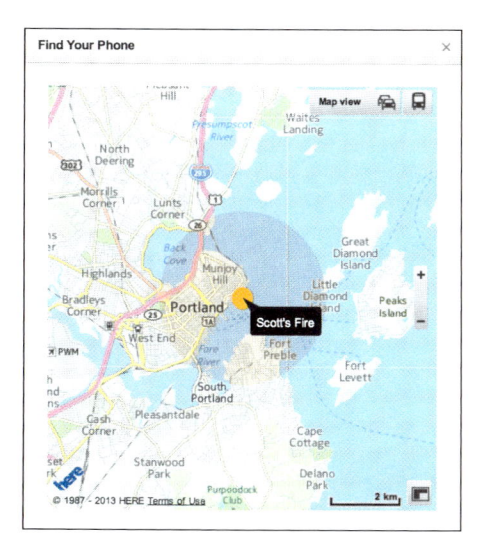

17.11 If your Fire is on and can connect to a network, Amazon.com can tell you where it is.

Remote Lock, Factory Reset, Remote Alarm

There are three more things you can do remotely in the Device Actions menu:

- Remote Lock is great if you haven't set a password on your Fire (though you really should). You can set a new PIN or password and display a message on the lock screen (17.12). If you've lost your phone, you can include your contact information in the message in the hopes that someone nice found it (17.13).

- Remote Factory Reset wipes all the data on the phone, returning it to factory defaults (the link to your Amazon account is also deleted).

- Remote Alarm causes your Fire to play an alarm sound for two minutes, or until the alarm is dismissed on the phone. This is great for locating a Fire that slipped between the sofa cushions.

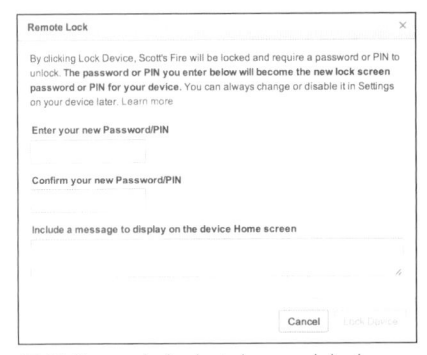

17.12 Remote-lock a lost phone and display a message.

17.13 The message on a remotely locked Fire.

CHAPTER 18

Help

Despite my best efforts with this book, I'm sure I haven't covered every situation that might crop up. You can get additional help with your Fire in three ways: Mayday, Self Service, and Amazon.com.

Mayday

One of the headline features of the Fire is Mayday. The Mayday button, in Quick Settings, connects you with an Amazon support person with just two taps (18.1).

The support person can see your phone's screen and hear you, and you can see and hear them. They cannot, however, see you, so there's no need to put on pants when using Mayday.

18.1 The Mayday button is in Quick Settings.

The Amazon representative can take over your phone and set things for you (with your permission) or even draw on your phone's screen to point out a setting or app feature.

To launch Mayday:

1. Swipe down to open Quick Settings, and tap the Mayday button.

2. Mayday checks the status of the Mayday service. If it's available, a Connect button appears (18.2). Tap it to connect with an Amazon representative.

3. A small box in the lower-right corner appears, with a video feed of the rep you're talking to (18.3). Tell them your problem, and they will try to help! Tap the Mute button to mute your end of the call, and tap End to end the call.

To turn on and off Mayday or to limit it to connecting over Wi-Fi only, open the left panel and tap Mayday Settings (18.4).

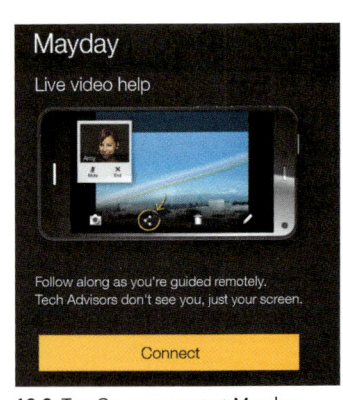

18.2 Tap Connect to start Mayday.

18.3 An Amazon representative appears. (I've blurred her face for privacy's sake.)

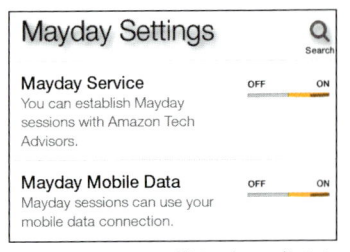

18.4 You can turn off Mayday or limit it to Wi-Fi only.

Self Service

You might not be in the mood to actually talk to someone about an issue. The Fire has built-in documentation that can answer some questions. To access it, launch Mayday and scroll down to More Options. Tap that to see the Self Service list (18.5). Search for a topic by tapping the search icon, or tap one of the entries:

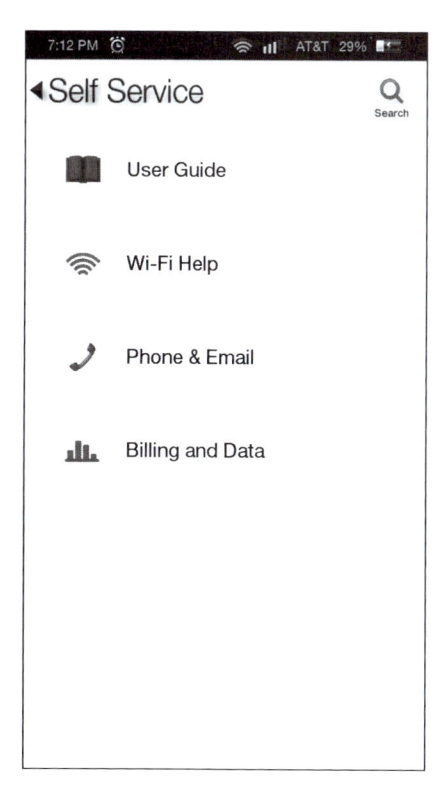

18.5 The Self Service option on your Fire is available when you don't want to talk to someone.

- User Guide has some basic user information about the Fire.
- Wi-Fi Help lists tips and tricks for dealing with wireless issues.
- Phone & Email has a link to Amazon's customer service.
- Billing and Data launches the AT&T app, which you need to log in to with your AT&T credentials.

Amazon.com help

Amazon has created a help site for the Fire on their website (18.6). The site includes searchable articles, how-tos, and videos that illustrate particular Fire functionality. Go to www.amazon.com/firephonesupport.

18.6 Amazon.com has support documents for your Fire.

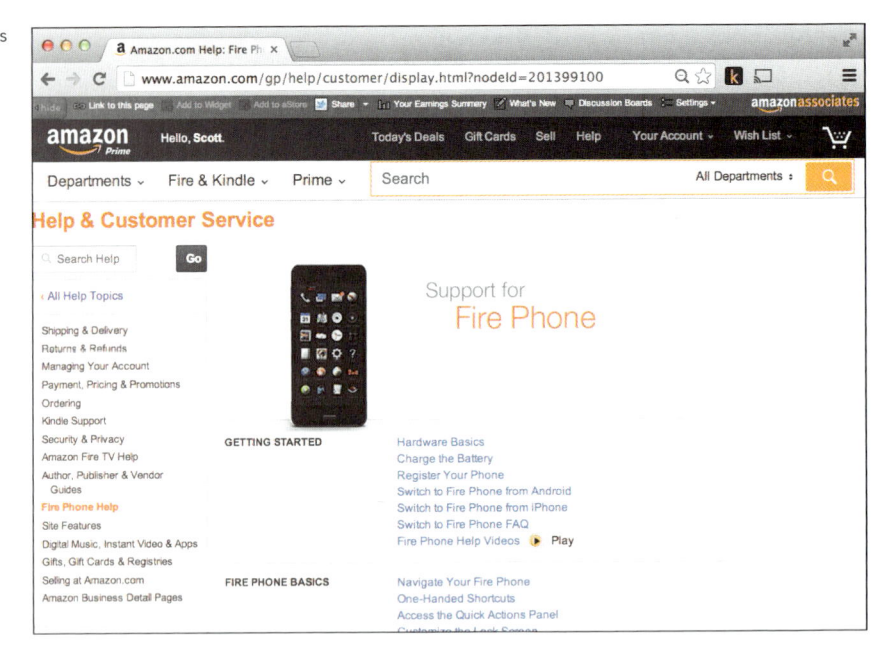

Index

Number
13 MP rear-facing camera, 8

A
address book. *See* Contacts app
advanced keyboard
   alternate characters, 33
   enabling, 33–35
albums
   displaying, 163–164
   purchasing, 192
alert sounds, customizing for contacts, 86
alternate characters, 33
Amazon Appstore
   accessing, 132
   Amazon Coins, 135
   app descriptions, 132
   app details, 132
   in-app purchases, 138
   category listings, 140–141
   Customer Reviews, 133
   left panel, 140–141
   More section, 141
   navigating, 139
   Permission section, 134
   Product Details section, 134
   recommendation engine, 133
   return policy, 134
   settings, 146–147
   sideloading apps, 142–146
   Test Drive, 136–137
Amazon Coins
   buying, 138–139
   explained, 135
Amazon integration, 3
Amazon order status, 17
Amazon Prime
   annual cost, 205
   getting year of, 3
   Instant Video, 207
   Kindle Lending Library, 209
   music, 207–208
   receiving year of, 3
   sharing shipping, 206–207
Amazon.com
   Deregister option, 217
   help, 222
   Locate on Map option, 217
   Remote Alarm, 218
   Remote Factory Reset, 218
   Remote Lock, 218
Android system, 3
.apk file, obtaining, 143
apps
   accessing from home screen, 12
   arranging on home grid, 22–24
   availability, 3
   buying and installing, 135
   grouping into collections, 23
   launching from Carousel, 12
   naming collections, 24
   pinning, 13
   switching between, 26
   unpinning, 13
apps preinstalled
   Calculator, 155–156
   Clock, 153–154
   Games, 147–149
   Maps, 149–152
Appstore
   accessing, 132
   Amazon Coins, 135
   app descriptions, 132
   app details, 132
   in-app purchases, 138
   category listings, 140–141
   Customer Reviews, 133
   left panel, 140–141
   More section, 141
   navigating, 139
   Permission section, 134
   Product Details section, 134
   recommendation engine, 133
   return policy, 134
   settings, 146–147
   sideloading apps, 142–146
   Test Drive, 136–137
AT&T app, downloading, 67–68
AT&T services, 94
attachments
   adding to email, 58
   listing, 52
   types for Docs app, 184
Attachments menu, 56–57
Audible trial membership, 11
Auto button, 21

B
backups, enabling, 11
banner notifications, toggling, 63–64
Bluetooth feature, 19
bookmarks
   adding, 115, 122
   displaying, 121
   editing, 122–123
   removing, 122
books
   download progress, 181
   downloading, 173
   jumping to chapters, 178
   navigating, 177
   opening, 175
   purchasing from Kindle Store, 181
brightness, adjusting, 21
browser data, clearing, 128. *See also* Silk browser
browsing history, searching, 111
buttons
   Camera, 6
   Home, 6
   Power, 6–7
   Volume, 6

INDEX   223

C

Calculator app, 155–156
calendar
  day view, 98–100
  dismissing notifications, 108
  events, 100–106
  launching, 97
  list view, 98–99
  month view, 98–99
  notification options, 108
  notifications, 107–108
  reminders, 108
  scheduled events, 98
  settings, 107–108
  syncing, 98
  today view, 97
  viewing, 97–100
calendar syncing
  Facebook events, 96
  turning off, 96
call forwarding, 94
call waiting, 93
caller ID, 94
calls. *See also* conference calls; Fire phone
  adding custom messages, 91
  adding labels, 89
  answering, 90
  checking calendar, 89
  keypad, 89
  making, 88–92
  making with voice controls, 202
  muting, 89
  Recent Calls button, 90
  speaker phone, 89
  surfing web, 89
  taking notes, 89
  viewing contacts, 89
  viewing time spent on, 89
Camera app
  Camera selector button, 158
  Flash button, 158
  Settings button, 158
  shooting video, 162
  Shutter button, 158
  starting, 157
  Still/Video button, 158
  taking pictures, 158–162
  Your Photos button, 158
Camera button, 6
capturing
  photos, 57
  videos, 57
Carousel. *See also* home screen
  launching apps from, 12
  Pin to Front, 13
  Pin to Home Grid, 14
  removing items from, 13
chapters, jumping to, 178
characters, alternate, 33. *See also* text; words
Clock app, 153–154
closed captions, displaying, 130
cloud features, 127–128
collections
  grouping apps into, 23
  moving, 24
  naming, 24
  opening, 24
  removing items from, 24
  using with Kindle app, 174
.COM button, displaying, 28
conference calls, 90. *See also* calls
Connected indicator, 10
contacts
  adding, 72
  customizing alert sounds, 86
  disabling syncing, 77
  navigating, 72–73
  searching, 72–73
  sorting, 78
  storing, 77
  transferring, 67–68
  verifying syncing, 66
  viewing during calls, 89
  VIPs, 50, 73–76
Contacts app. *See also* email addresses
  adding labels, 69
  adding picture to profile, 69
  Address section, 70
  copying info to Clipboard, 70
  custom ringtones, 72
  Delete button, 70
  displaying fields, 68
  Edit button, 70
  Email field, 69
  entering name, 69
  joining, 75–76
  launching, 68
  menu icon, 70
  navigational options, 66
  Peek feature, 71
  personal profile, 68
  Phone field, 69
  quick view, 74
  saving entries, 70
  sending address to Maps app, 70
  settings, 76–78
  Share button, 70
  source, 70
  splitting, 75–76
  taking actions, 70
contacts list, returning to, 71
conversation settings, 62–63
conversations, deleting messages in, 81
copying text messages, 84
cursor, inserting, 37–38

D

day view, displaying for calendar, 98–100
default settings, restoring, 130
deleting
  email accounts, 44–45
  email messages, 53, 58
  events in calendar, 101
  photos, 165
  text messages, 81, 84
deregistering phones, 217
dictating messages, 32–33
dictionary
  adding words to, 29–30
  editing words in, 30

Docs app
    attachment types, 184
    launching, 184
documents, getting onto Fire, 184
double tapping, 4
Dynamic Perspective
    features, 2
    sensors, 6

E

email accounts
    adding attachments to, 58
    deleting, 44–45
    detecting, 43
    display settings, 46
    Exchange, 39, 42
    IMAP, 39, 44
    jumping to, 51
    POP, 39, 44
    senders, 52
    setting up, 40–44
    SMTP server details, 44
    sync and data settings, 47–49
    VIPs, 50
email addresses. *See also* Contacts app
    adding subjects to, 56
    removing, 56
    removing from recipient fields, 56
    verifying, 56
Email button, using with calendar, 101–102
email messages. *See also* inbox
    adding VIPs, 56
    appending signatures to, 48
    Archive or Delete button, 53
    attachments, 52, 56–57, 61
    banner notifications, 63–64
    changing font size, 60
    changing text size, 60
    color coding, 51
    composing, 55–58
    conversation settings, 62–63
    customizing, 60–64
    Delete button, 53
    deleting, 44, 58
    displaying from single accounts, 50
    displaying recent, 54
    Exchange, 44
    flagging, 59
    going to, 54
    including originals in replies, 61
    left panel, 51
    mass-editing, 59
    Menu button, 53
    moving, 58–59
    notifications, 64
    previewing inboxes, 50
    previewing on Carousel, 50
    receiving notifications, 63–64
    recent, 50
    Respond button, 53
    scrolling through, 51
    searching, 51
    sending, 55
    Show Complete Message button, 53
    Show Embedded Images, 61
    sound settings, 64
    subjects, 53
    updating search results, 51
    viewing recent, 50
    voice controls, 203–204
emoticons, 35
encryption, 213–214
ESV Prompt, 128
events in calendar
    accessing details, 100–101
    Account, 105
    All Day, 104
    creating, 104–106
    Delete button, 101
    Edit button, 102
    Email button, 101–102
    Ends, 104
    with guests, 103
    Guests, 105
    More section, 101
    no guests, 103
    Notes, 106
    in the past, 103
    receiving invitations, 102
    Reminders, 105
    Repeat, 104
    responding to invitations, 106
    Starts, 104
    Title, 104
    Where, 105
Exchange email accounts, 39, 42, 44

F

Facebook, syncing events, 96
Facebook accounts, connecting to, 11
factory defaults, restoring, 130
Find Your Phone feature, 217
finding text messages, 85. *See also* searching
Fire phone. *See also* calls
    back, 8
    deregistering, 217
    front, 6
    locating on map, 217
    registering, 9
    setting up, 9–11
    turning off, 7
    turning on, 9
Firefly
    enabled apps, 169
    features, 2, 166–167
    launching, 6
    left panel, 169
    movies, 168
    pointing at URLs, 166–167
    searching history, 169
    Sharing panel, 170
    songs, 168
    starting, 166
    TV shows, 168
Flashlight feature, 19
flinging video, 198–199
font size, changing for email, 60
forwarding text messages, 84

G

Games app, 147–149
Gmail account
   adding labels, 54
   reauthorizing, 49
   setting up, 40–44
Google's Android system, 3
Google's App Store, 4
GPS, enabling, 10

H

HDR (high dynamic range), 162
headphone jack, 7
headphones, using with ringer, 93
help. See Mayday
Home button, 6
home grid
   accessing, 22
   appearance, 21
   arranging apps, 22–24
   Cloud tab, 22
   Device tab, 22
   installed apps, 22
   moving icons, 23
   pinning to, 25
   swiping, 22
   switching pages, 22
home screen. See also Carousel
   Carousel, 12
   left panel, 15–16
   right panel, 16–17

I

icons, moving on home grid, 23
IMAP email accounts, 39, 44
importing contacts, 67–68
inbox. See also email messages
   accessing, 51
   getting back to, 54
   previewing, 50
installing apps, 135
Instant Video app. See also movies; TV shows; videos
   buying videos, 196
   flinging video, 198–199
   links, 194
   renting videos, 195–196
   searching store, 195
   suggestions, 195
   using, 207
   viewing shows and movies, 194
   watchlist, 199
   X-Ray button, 198
invitations to events
   receiving, 102
   responding to, 106

K

keyboards
   advanced, 33–35
   alternatives, 34–35
   emoticons, 35
   entering numbers, 28
   entering symbols, 28
   number pad, 34
   text selection, 34–35
keypad, accessing for calls, 89
Kindle app
   book options, 174
   bookmarks, 178–179
   Cloud library, 172
   collections, 174
   Device library, 172
   downloading books, 173
   filtering options, 173
   highlights, 178–179
   jumping to chapters, 178
   left panel, 178
   libraries, 172
   List view, 173
   lookups, 180
   More settings, 176
   navigating books, 177
   notes, 178–179
   Notes and Marks, 177
   opening books, 175
   popular highlights, 176
   progress options, 175
   reading screen, 175–178
   restoring menus, 176
   Search icon, 172
   Share, 177
   sharing passages, 179
   slider, 177
   Store icon, 172
   translation, 180
   X-Ray, 176–177
Kindle First, 210
Kindle Lending Library, 209
Kindle Store
   book listings, 180
   Download Sample button, 180
   purchasing books, 181
   shopping in, 180

L

language, selecting, 9
LED flash, 8
lenticular photography, 160–161
library. See Kindle Lending Library
lock icon, 19
lock screen, setting, 212
long tapping, 4

M

magazine content
   Subscribe Now button, 183
   viewing, 182–183
mail. See email accounts
Mail app, tapping, 51. See also messages
Maps app, 149–152
Mayday
   built-in documentation, 221
   features, 2, 20
   launching, 220
   Self Service, 221
   turning off, 220
message list, returning to, 81
messages. See also inbox; Mail app
   adding VIPs, 56
   appending signatures to, 48
   Archive or Delete button, 53

attachments, 52, 56–57, 61
banner notifications, 63–64
changing font size, 60
changing text size, 60
color coding, 51
composing, 55–58
conversation settings, 62–63
customizing, 60–64
Delete button, 53
deleting, 44, 58
displaying from single
  accounts, 50
displaying recent, 54
Exchange, 44
flagging, 59
going to, 54
including originals in replies, 61
left panel, 51
mass-editing, 59
Menu button, 53
moving, 58–59
notifications, 64
previewing inboxes, 50
previewing on Carousel, 50
receiving notifications, 63–64
recent, 50
Respond button, 53
scrolling through, 51
searching, 51
sending, 55
Show Complete Message
  button, 53
Show Embedded Images, 61
sound settings, 64
subjects, 53
updating search results, 51
viewing recent, 50
voice controls, 203–204
Messaging app. *See also* text
  messages
  opening, 79
  right panel, 81
micro USB port, 8
microphone, 7
month view, displaying for
  calendar, 98–100

movies, identifying with Firefly, 168.
  See also Instant Video app
MMS messages, 80
multitasking, 26
music
  managing on playlists, 190–191
  pausing, 188
  purchasing, 191–192
Music app. *See also* songs
  albums, 186
  artists, 186
  genres, 186
  launching, 186
  navigating songs, 187
  play controls, 187
  playlists, 186, 188–191
  Prime, 207–208
  repeating songs, 187
  songs, 186
muting calls, 89

N
Newsstand app, 182–183
notifications
  customizing for text messages,
    85–86
  turning on, 64
number pad keyboard, 34
numbers, entering, 28

P
pages
  bookmarking, 115
  returning to, 116
  saving, 123
  sharing, 115
panels
  dismissing, 15
  features, 14–15
  home screen left, 15–16
  home screen right, 16–17
  Quick Settings, 17–21
  revealing, 15
panoramas, taking, 159–160

password
  changing for voicemail, 94
  protection, 19
  using, 212
Peek feature
  overview, 14
  using with Contacts app, 71
period, inserting, 28
phone calls. *See also* Fire phone
  adding custom messages, 91
  adding labels, 89
  answering, 90
  checking calendar, 89
  keypad, 89
  making, 88–92
  making with voice controls, 202
  muting, 89
  Recent Calls button, 90
  speaker phone, 89
  surfing web, 89
  taking notes, 89
  viewing contacts, 89
  viewing time spent on, 89
photo albums, displaying, 163–164
photo storage, 2
photography, lenticular, 160–161
photos. *See also* pictures
  adding to profile, 69
  capturing, 57
  deleting, 165
  displaying info about, 165
  long tapping, 165
  sharing, 165
pictures. *See also* photos
  adding to profile, 69
  HDR (high dynamic range), 162
  lenticular photography, 160–161
  panoramas, 159–160
  rapid succession, 158
  saving to camera roll, 163
  sending in text messages, 83
  taking, 158
PIN (personal identification
  number), 212
Pin to Front, 13
Pin to Home Grid, 14

INDEX   227

pinching, 4
pinning to home grid, 25
playlists
　adding songs to, 188–189
　creating, 190
　downloading songs on, 190
　editing, 189
　managing music, 190–191
　removing songs from, 189
　renaming, 189
　verifying, 188
POP email accounts, 39
Power button, 6–7
predictive text. *See also* text
　dictionary additions, 29–30
　suggestion bar, 29
Prime
　annual cost, 205
　getting year of, 3
　Instant Video, 207
　Kindle Lending Library, 209
　music, 207–208
　receiving year of, 3
　sharing shipping, 206–207
profile, adding picture to, 69

Q

Quick Settings
　accessing, 18
　Airplane Mode, 18
　Bluetooth, 19
　dismissing, 18
　Flashlight, 19
　Mayday, 20
　opening, 4
　Search, 20
　Settings, 19
　Sync, 19
　using Peek, 18
　Wi-Fi, 18
Quick Switch, 26

R

Reader mode, 117–118
registering Fire phone, 9–10
Remote options
　Alarm, 218
　Factory Reset, 218
　Lock, 218
renting videos, 195–196
Restart option, 7
ringer volume, setting, 93
ringtones
　controls, 92
　customizing for contacts, 72
　setting, 92

S

scanner. *See* Firefly
screen, high-definition, 6
Search feature, 20
search field, revealing, 51
searching. *See also* finding text
　　messages; web searches
　contacts, 72–73
　via voice controls, 204
security
　Amazon.com, 216
　Deregister option, 217
　Device Actions, 218
　encryption, 213–214
　password, 212
　PIN (personal identification
　　number), 212
　Remote Alarm, 218
　remote Device Actions, 218
　Remote Factory Reset, 218
　Remote Lock, 218
　setting lock screen, 212
　updates, 215
　warnings, 130
selecting text, 35–38
Self Service option, 221
server settings, accessing, 49

settings. *See* Quick Settings
Settings feature, 19
sharing photos, 165
shortcuts, creating on home grid, 25
sideloading apps, 142–146
signatures, appending to email, 48
Silk browser. *See also* browser
　　data; websites
　creating tabs, 119
　features, 110
　Find in Page option, 115–116
　going to websites, 111
　launching, 110
　links, 113–114
　magnifying glass icon, 111
　menu button, 115
　most visited pages, 110
　open tabs, 116, 119
　opening tabs, 120
　Reader mode, 117–118
　recently visited websites, 111
　Request Another View, 116
　scrolling pages, 112
　scrolling through tabs, 120
　search results, 111–112
　site links, 117
　suggestions, 111
　tab options, 120
　web browsing history, 111
　web searches, 111
Silk navigation
　bookmarks, 121–123
　Cloud Features, 127–128
　Downloads section, 125
　History panel, 126
　Saved Pages, 123
　Settings panel, 126–130
　Trending Now, 124
　Your Data section, 128–129
SIM tray, 8
site links, displaying, 117
SMS messages, 80
SMTP server details, entering, 44

songs. *See also* Music app
   adding to playlists, 188–189
   identifying with Firefly, 168
   lyrics link, 187
   pausing, 188
   playing from Cloud library, 187
   previewing, 192
   removing from playlists, 189
sounds
   customizing for contacts, 86
   setting for email notification, 64
speaker phone, 89
speakers, 7
spellcheck, 31
status bar, hiding and displaying, 14
subjects, adding to email messages, 56
swiping, 4
symbols, entering, 28
sync and data settings
   appending signatures, 48
   Automatic, 48
   Days to Sync, 48
   explained, 47
   Manual, 48
   Server Settings, 49
Sync feature, 19

T

tabs
   creating in Silk, 119
   opening in Silk, 120
   scrolling through, 120
tapping, 4
Test Drive apps, 136–137
text. *See also* characters; predictive text; words
   inserting cursor, 37–38
   selecting, 35–38
   trace typing, 31–32
text display, changing in Reader mode, 118
text messages. *See also* Messaging app
   conversations, 80
   copying, 84
   creating, 82–83
   deleting, 81, 84
   dictating, 83
   finding, 85
   forwarding, 84
   message list, 80
   MMS, 80
   notifications, 85–86
   reading, 84–86
   receiving attachments, 80–81
   sending, 82–84
   sending pictures, 83
   sending videos, 83
   SMS, 80
   unread messages, 79–80
   voice controls, 203–204
text selection keyboard, 34–35
text size, changing for email, 60
text to speech, 32–33
tilting, 4
touch device
   double tapping, 4
   Fire as, 4
   long tapping, 4
   pinching, 4
   swiping, 4
   tapping, 4
   tilting, 4
trace typing, 31
transcription, 32–33
transferring
   sideloaded apps, 144–146
   videos, 199
Trending Now, 124
turning off Fire phone, 7
turning on Fire phone, 9
tutorial, 11
TV shows. *See also* Instant Video app
   identifying with Firefly, 168
   searching with Instant Video, 195
Twitter accounts, connecting to, 11
typing, 31

U

updates, receiving, 215
URL/search box, viewing, 114
USB, using to sideload apps, 144–146
USB port, micro, 8

V

VCF file, exporting contacts to, 67–68
Vibrate option, 64
vibration, enabling and disabling, 92
video settings
   Clear Video Search History, 200
   Dolby Digital Plus 5.1, 200
   HD Download Quality, 200
   mobile network, 200
   SD Download Quality, 200
videos. *See also* Instant Video app
   buying, 196
   captions, 197
   capturing, 57
   displaying, 164
   Exit Playback, 197
   menu options, 197
   Next Up button, 197
   Pause/Play button, 196
   Play on control, 197
   playing, 165
   renting, 195–196
   sending in text messages, 83
   shooting, 162
   Skip Back button, 197
   timeline, 197
   transferring, 199
   Volume control, 197
   Watch Now, 196
   watching, 196–199
   Zoom control, 197
VIPs
   adding to email, 56
   assigning contacts as, 73–76
   designating contacts as, 50

voice controls
    messaging, 203–204
    phone calls, 202
    searching, 204
Voicemail Password, 94
volume, raising and lowering, 6, 93

W

watchlist for videos, accessing, 199
web searches, performing, 111.
    See also searching

webpages
    bookmarking, 115
    returning to, 116
    saving, 123
    sharing, 115
websites. *See also* Silk browser
    downloading, 115
    going to, 111
    long tapping images on, 114
    searching, 115–116
    sharing pages, 115
    viewing URL/search box, 114

Wi-Fi
    adding networks, 10
    downloading attachments, 61
    turning on in Airplane mode, 19
words. *See also* characters; text
    adding to dictionary, 29–30
    editing in dictionary, 30
    selecting, 35
    spellcheck, 31
    suggestion bar, 29